# Quantum Radio Frequency Physics

L. D. STEPIN

# Quantum Radio Frequency Physics

TRANSLATED BY SCRIPTA TECHNICA INC.
EDITED BY HENRY H. STROKE

 The M.I.T. Press
*Massachusetts Institute of Technology*
*Cambridge, Massachusetts*

Lev Dmitriyevich Stepin
Kurs Lektsiy po Kvantovoy Radiofizike
*Published originally by the A. M. Gor'kiy State University of
Khar'kov, 1963, under the title* Lectures in Quantitative Radiophysics

*Library of Congress Catalog Card Number: 65-25207
Printed in the United States of America*

## Translation Editor's Preface

The recent upsurge in research on the interactions of electromagnetic radiation with matter, particularly in the field of masers and lasers, has stimulated renewed interest in the several basic resonance systems. The current advances in these fields are much too rapid to permit a complete and exhaustive presentation, but the present work should serve as a useful guide and reference for scientists who wish to pursue studies in this area.

The translation has kept the author's practical units for the magnetic field in the few places where it occurs. For a qualitative feeling a conversion factor of about 10 will be adequate (1 kiloampere turn/meter corresponds to $4\pi$ gauss).

H. H. Stroke

New York, March 1965

# Contents

# Foreword

Quantum radio frequency physics, which encompasses phenomena of resonant interaction between microwaves and radiofrequency waves and matter, is a relatively young branch of physics. Electron paramagnetic resonance was discovered in 1944 and nuclear magnetic resonance in 1946 and the first quantum oscillators and amplifiers were built approximately ten years ago. At present, quantum radio frequency physics has already yielded a number of important results for practical applications. Radio frequency spectroscopic methods of investigation are a powerful tool in the hands of physicists, chemists, biologists and other researchers. Development of low noise quantum amplifers has opened new possibilities in radio astronomy, radar and communications.

A considerable number of articles on special topics of quantum radio frequency physics is available. As a rule, these works are highly specialized and are aimed at the reader who is well versed in the subject. At the same time a shortage of textbooks on the subject is keenly felt.

This work presents a series of lectures delivered at the Department of Radiophysics of the Kharkov State University for students who do not specialize in quantum radio frequency physics. The main purpose of the course is to familiarize the student with both the physical concepts and some of the theoretical problems which form the foundation of quantum radio frequency physics.

The book discusses the phenomena of electron and nuclear magnetic resonance and the methods used in their observation. The operation of two-level and three-level masers is described. The last chapters of the book are devoted to radio frequency spectroscopy with gases and to the operation of a molecular oscillator using a beam of ammonia molecules.

The author expresses his gratitude to Assistant Professor V. K. Tkach, I. N. Komar and to other staff members of the Department of Radio Spectroscopy at the Kharkov State University for their valuable advice and remarks.

*Chapter I*

# Introduction

## 1. QUANTUM RADIO FREQUENCY PHYSICS

Quantum radio frequency physics is one of the most intensively developing branches of science at the present. It is based on the phenomenon of resonant interaction with matter of electromagnetic radiation in the microwave and RF regions. As a result of this interaction, a quantum of electromagnetic energy is either radiated or absorbed. Every atom or molecule can absorb or radiate electromagnetic waves of various wavelengths. The complete set of such waves constitutes the spectrum of the atom or molecule. The study of atomic and molecular spectra and the determination of their energy levels is called spectroscopy. The development of spectroscopy started with the study of spectra in the visible region and then progressed to invisible waves—ultraviolet, x rays, infrared—and finally to those regions of the electromagnetic spectrum that can be observed by radio frequency techniques. Studies of spectra in the microwave and RF regions fall within the domain of radio

frequency spectroscopy, which is one of the branches of quantum
radio frequency physics.

Radio frequency spectroscopic methods can be used to study
matter in its solid, liquid or gaseous state. The study of gases
by radio frequency spectroscopic methods is based on the following considerations. It is known that the energy levels of
polyatomic molecule are primarily determined by the energy
states of the atoms that form the molecule. Moreover, the
energy levels depend on vibrations of the atoms with respect to
each other as well as on the rotation of the entire molecule.
Figure 1 is a schematic representation of the vibrational-
rotational level structure of such a molecule. Each electronic
level of the molecule is split into a number of sublevels which
correspond to various vibrational states, characterized by
quantum numbers $v_i$. The vibrational levels in turn have a

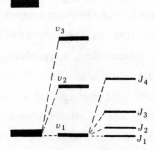

FIG. 1. Vibrational-rotational
structure of energy levels

number of sublevels which correspond to possible rotational
states and are characterized by quantum numbers $J_k$. The
transitions between the vibrational levels are studied by infrared spectroscopy. The transitions between the rotational
levels, however, for a number of cases fall within the region of

microwave spectroscopy. For example, in a molecule of cyanogen bromide, CNBr, the first three transitions are observed at frequencies of 8226 Mc/s, 16,452 Mc/s and 24,676 Mc/s. The total number of lines in the microwave region is 37. The apparatus shown schematically in Fig. 2 can be used for observing rotational transitions. It consists of a microwave generator G, an absorption cell AC and a detector D. These components are fundamental for any gaseous radio frequency spectrometer. Microwave radiation goes through the absorption cell, where it interacts with the molecules of the gas under investigation. If the frequency of the incident radiation equals the frequency of the transition under study, energy is absorbed and the output power, which is registered by the detector, decreases.

FIG. 2. Block diagram of a simple gaseous radio frequency spectrometer

The spectral lines observed in a gas, when it is transformed to a liquid or solid, become very wide and weak and practically cannot be observed. This is because in a condensed medium the interaction between molecules is very strong, causing each energy level to be smeared out into a band. Thermal motion also causes broadening of the energy levels. Therefore, instead of having absorption of radiation at a sharp frequency, one has absorption of radiation with a more or less arbitrary spread of frequencies. This renders radio frequency spectroscopic investigation of solids and liquids rather difficult. Even in the solid state, however, one can observe discrete energy levels resulting from magnetic properties of atoms and molecules. If an atom or ion has a magnetic moment $\vec{M}$ and is located in a constant magnetic field, the moment $\vec{M}$ can assume a number of discrete directions with respect to the direction of the external magnetic field (see Fig. 4). For every orientation of the magnetic moment there is a corresponding energy of interaction of the atom (or ion) with the magnetic field. As a consequence of this, each atomic level is split into a number of sublevels with the magnitude of the splitting (and hence the transition frequency) depending on the strength of the external magnetic field. This phenomenon is known as the Zeeman effect. Transitions between the Zeeman-split components can be easily observed by radio frequency spectroscopic methods. A new element of the radio frequency spectrometer used to observe such transitions that we did not have in the gaseous radio frequency spectrometer is the external magnetic field applied to the investigated sample. Since in radio frequency spectroscopy of liquids and solids one observes a reorientation of the magnetic moment when the generator frequency and the transition frequency become equal, this phenomenon is called electron paramagnetic resonance if the magnetic moment pertains to an atom or an ion, and nuclear magnetic resonance if the magnetic moment is that of the nucleus of an atom.

Until now, the discussion was limited to absorption of microwaves and RF waves by matter and corresponding transitions of atoms or molecules from lower energy states to higher energy states. It is entirely possible to formulate the reverse concept: to use atoms and molecules not for absorption but for emission of microwave radiation. This problem was formulated and its solution resulted in the creation of new methods for generation and amplification of microwaves and RF waves, and in the invention of quantum oscillators and amplifiers and of quantum radio engineering. Quantum radio engineering is a second very important branch of quantum radio frequency physics.

In general, the principle of quantum oscillators and amplifiers is as follows. Some method is used to bring a sufficiently large number of atoms or molecules from a lower energy state to a higher. These atoms or molecules decay to their ground state, radiating monochromatic electromagnetic waves. If the atoms or molecules in the higher energy state are continuously replenished, a source of microwave radiation is obtained.

From the preceding paragraphs it is evident that quantum radio frequency physics is a science which deals with absorption and emission spectra of microwaves and RF waves and utilizes them either for studies of atoms or molecules or for the development of new methods of generation and amplication of electromagnetic radiation in the microwave region.

## 2. A BRIEF HISTORICAL SURVEY OF THE DEVELOPMENT OF QUANTUM RADIO FREQUENCY PHYSICS

Studies of interactions between radiation and matter began more than a half a century ago. In 1913, V. K. Arkad'yev observed selective absorption of radio waves by ferromagnetic materials [1]. The work of Arkad'yev was not spectroscopic in character because his results were not interpreted in terms of a spectrum. Nine years after Arkad'yev's tests, Einstein and Ehrenfest showed theoretically that a change in orientation of atomic dipole moments in a magnetic field should be accompanied by radiation or absorption of electromagnetic waves [2].

This was the prediction of the phenomenon of magnetic resonance. During the next year (1923) J. Dorfmann gave a quantum-mechanical explanation for Arkad'yev's results and predicted the existence of the "photomagnetic effect," i.e., changes in magnetic substates of paramagnetic substances under the influence of RF radiation [3].

In 1934 Cleeton and Williams carried out the first investigation of the microwave spectra of gases: they investigated absorption in ammonia vapor and discovered a wide absorption band, centered at a wavelength of about 1.25 cm [4]. The spectrometer of Cleeton and Williams was far removed from contemporary radio frequency spectrometers. It was essentially a hybrid of an optical and a radio frequency spectrometer. A split-anode magnetron was used as a generator and the microwave radiation, focused with a brass parabolic mirror, was transmitted through a rubber bag containing ammonia at atmospheric pressure. An optical system was used to measure the wavelength. There was no electronic amplification.

In 1935 L. D. Landau and Ye. M. Lifshits showed theoretically the existence of magnetic resonant absorption in ferromagnetic materials [5]. In 1936 Gorter started a systematic investigation of the interaction between radiation and paramagnetic substances [6]. In 1937 Rabi developed the theory of magnetic resonance for particles with a magnetic moment and applied this theory to the experimental determination of nuclear magnetic moments. It should be noted that in his experiments Rabi used atomic and molecular beams [7].

All this work was the forerunner of the birth of a new scientific field, the study of absorption spectra of electromagnetic waves in the microwave and RF regions. In essence, these works originated radio frequency spectroscopy since they supplied the fundamental concepts on which radio frequency spectroscopy was built. It must also be noted that, even in those early efforts, three fundamental branches of radio frequency spectroscopy can be clearly discerned.

The development of the first branch was initially more vigorous than that of the other branches and used Rabi's method of observation of magnetic resonance in atomic and molecular beams. In 1947 these efforts resulted in two major discoveries: the establishment of the shift of $S$ levels in hydrogenlike atoms and the detection of the anomalous magnetic moment of the electron. The investigations were conducted to verify and refine the Dirac theory. The basic achievement of the relativistic theory of Dirac was that it gave a "correct" fine structure of the energy levels and showed that an electron should possess an angular momentum equal to ½ in units of $h/2\pi$ ($h$ = Planck's constant) and that this angular momentum is related to the magnetic moment, which equals one Bohr magneton. Numerous investigations of the energy levels of the hydrogen atom, however, discovered a slight discrepancy between the experimental data and the calculations based on the Dirac theory. This was first observed in 1934, but the optical techniques used then were not sufficiently precise to permit a definite conclusion as to the existence of any deviation from the Dirac theory. This situation continued until 1947 when Lamb and Retherford used a radio frequency spectroscopic method instead of the optical method.

According to the Dirac theory, the state $2^2S_{1/2}$ has the same energy as the state $2^2P_{1/2}$ (Fig. 3). This result was subject to verification. The method used in the experiment was relatively simple. A beam of hydrogen atoms was bombarded by a beam of electrons. By this means one atom in a million was excited from the ground state $1^2S_{1/2}$ to the $2^2S_{1/2}$ state, which is a metastable state (transitions to the ground state $1^2S_{1/2}$ are forbidden by the selection rule $\Delta L = \pm 1$). The beam of atoms was incident on a metallic detector from which the metastable atoms "ejected" electrons, producing current flow which was measured at the detector. On their way to the detector the atoms

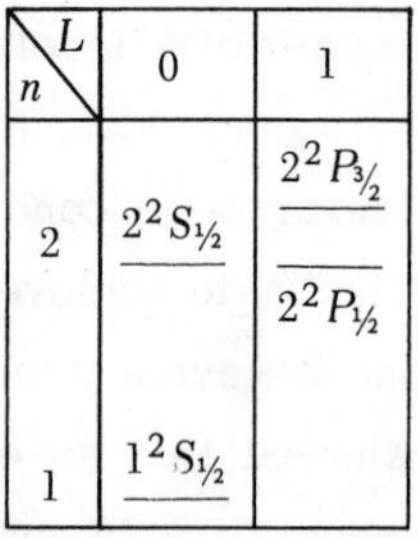

| $\diagdown L$  $n$ | 0 | 1 |
|---|---|---|
| 2 | $2^2S_{1/2}$ | $2^2P_{3/2}$ ——— $2^2P_{1/2}$ |
| 1 | $1^2S_{1/2}$ | |

FIG. 3. Energy levels of the hydrogen atom according to the Dirac theory

passed through a region of constant magnetic field. Each energy level was thus split according to the possible orientations of the total angular momentum of the atom $\vec{J} = \vec{L} + \vec{S}$, the number of sublevels being $2J + 1$. The splitting $\Delta E$ is proportional to the magnetic field $H$. In addition, the beam of atoms was subjected to the action of an RF field, the energy quantum $h\nu$ of which corresponded to the difference in energies between one of the Zeeman components of the state $2^2S_{1/2}$ and any component of $2^2P_{1/2}$ or $2^2P_{3/2}$. Then there were induced transitions from the metastable $2S$ state to the $2P$ state, in which the atom has a lifetime of $1.6 \cdot 10^{-9}$ sec, and as a result it decayed to its ground $1S$ state. Due to transitions of this nature the number of metastable atoms was immediately reduced, causing a drop in the detector current. From this is easily determined the resonant frequencies and also the difference between the $2S$ and $2P$ levels. It was shown that the $2^2 S_{1/2}$ level is 1062 Mc/s higher than the $2^2 P_{1/2}$ level.

An analogous displacement of $S$ levels, which was subsequently called the Lamb shift, was also detected in deuterium and helium atoms. Thus it was finally established that the Dirac theory does not give a complete description of a hydrogenlike atom and that some details implied by that theory are questionable.

The application of Rabi's radio frequency spectroscopic method to the study of the hyperfine structure of hydrogen and deuterium atoms (work of Naffe, Nelson and Rabi [8]) also uncovered discrepancies between the experimental results and the values computed from the theory. It was soon noted that these discrepancies can be easily explained if the magnetic moment $\mu_e$ of an electron is assumed to be somewhat greater than a Bohr magneton $\mu_0$, i.e.,

$$\mu_e = \mu_0 \left(1 + \frac{a}{2\pi}\right)$$

where $a = \dfrac{2\pi e^2}{hc} = 1/137$ is the fine-structure constant.

The discovery of the Lamb shift and of this correction to the magnetic moment of the electron strongly influenced the development of quantum electrodynamics and pointed out the immense value of the radio frequency spectroscopic technique. It was established later that both of the above discoveries were mainly due to Dirac's neglecting the interaction of the electron with vacuum fluctuations of the electromagnetic and electron-positron fields.

The second branch of radio frequency spectroscopy deals with studies of absorption in gases. In this field the early work was by Cleeton and Williams (1934), who observed absorption in ammonia. This effort, however, was isolated. Further development of gaseous radio frequency spectroscopy was hindered by the absence of necessary equipment. This equipment was designed during the Second World War. Its appearance was connected with the development of radar, which was helpful in the rapid development of this branch of radio frequency spectroscopy. This is substantiated by the number of published papers: before 1946 only five papers devoted to these problems were published, in 1946 there were 22 papers, and in 1947 already 50 papers. The first efforts in gaseous radio frequency spectroscopy were devoted to an explanation of the absorption of radio waves in the atmosphere. The work of the Soviet scientist V. L. Ginsburg in 1942 and subsequently the work of the American scientist Van Vleck showed that the absorption of centimeter radio waves in the atmosphere is caused mainly by water vapor. An extremely strong water vapor absorption exists for wavelengths from 1.2 to 1.6 cm. This prevented the application of the wavelength of 1.25 cm in radar, for which the USA had already developed klystrons, magnetrons and other components.

The development of gaseous radio frequency spectroscopy resulted in the molecular oscillator, whose feasibility was pointed out by N. G. Basov in his speech at the All-Union Conference on Radio Frequency Spectroscopy in 1952. Within a few years the first molecular oscillators, utilizing a beam of ammonia molecules, were designed in both the USA and the USSR. The idea behind the molecular oscillator is exceedingly simple: a beam of molecules in their excited state is transmitted through a cavity tuned to a frequency that is the saem as the frequency of the electromagnetic radiation that results when the excited molecules return to their ground state. If the energy radiated in the cavity exceeds the losses, the cavity becomes a generator of microwave power.

The third branch of radio frequency spectroscopy is connected with studies of the absorption of radiation by solids. A large amount of work in this field was performed by Gorter. He did not, however, detect the phenomenon of magnetic resonance. Magnetic resonance was first detected experimentally by Ye. K. Zavoyskiy in 1944 during his study of paramagnetic salts. The phenomenon was called electron paramagnetic resonance. The magnetic resonance caused by nuclear magnetic moments (so-called nuclear magnetic resonance) was first observed in 1946 by Purcell and Bloch. The theory of electron paramagnetic resonance was developed by Ya. I. Frenkel' and supplemented later by the work of S. A. Altschuler and B. M. Kozyrev. In 1948 R. P. Penrose and B. M. Kozyrev, as well as S. A. Altschuler and S. G. Salikhov, succeeded in the first experimental observation of the hyperfine structure in electron paramagnetic resonance. Further investigations of electron paramagnetic resonance resulted in the development of paramagnetic oscillators and amplifiers based on a principle which is analogous to that of molecular oscillators and amplifiers.*

The above excerpts from the history of radio frequency physics, although sketchy, illustrate some of the directions of stages of its growth.

---

*During the last years other branches of radio frequency spectroscopy of solids have been intensively investigated. These include diamagnetic (cyclotron) resonance in semiconductors, cyclotron resonance in metals, ferromagnetic resonance in ferromagnetic and antiferromagnetic substances, and combined resonance in semiconductors. Investigation of these phenomena at the present time is a powerful tool for studies of the energy spectra in solids.

*Chapter II*

# Fundamentals of Magnetic Resonance Theory

## 3. MAGNETIC RESONANCE. LONGITUDINAL AND TRANSVERSE RELAXATION

Let us examine an isolated spin which possesses a mechanical angular momentum (quantum number $F$) and a corresponding magnetic moment $\mu$, and placed in an external magnetic field $H$. We state without proof the well-known quantum-mechanical fact that the magnetic and mechanical momenta are related to each other by $\mu = g\mu_0 F$ (where $\mu_0$ is the electronic or nuclear magneton and $g$ is the gyromagnetic ratio or $g$ factor), and that the orientation of the magnetic moment is due to the external magnetic field. The projection of the magnetic moment on the direction of $H$ is $\mu' = m\mu/F$, where $m$ is the magnetic quantum number, which takes on discrete values $F, F - 1, \ldots -(F - 1), -F$ (a total of $2F + 1$ values). For each orientation of the magnetic moment there is an energy of interaction between the particle and the magnetic field, $E = -m\mu H/F$ (Fig. 4). The separation between two adjacent levels, which is the same for all levels, is

$$\Delta E = g\mu_0 H \tag{1}$$

Now, if the paramagnetic spin is subjected to a perturbation by an alternating magnetic field with resonant frequency $\nu_0$ such that the quantum $h\nu_0$ is exactly the same as the difference between the levels, $\Delta E$, and if the direction of the alternating field is perpendicular to the direction of the static magnetic field, then there will be induced transitions between adjacent energy

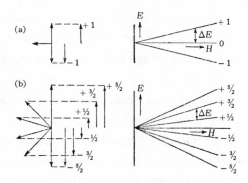

FIG. 4. Possible orientations of angular momenta with respect to the direction of the external magnetic field $H$ and their corresponding energy levels. a—$F = 1$; b—$F = 5/2$

levels (magnetic dipole transitions). Furthermore, the change in the magnetic quantum number (selection rule) will satisfy the condition $\Delta m = \pm 1$. Therefore the condition for magnetic resonance is

$$h\nu_0 = g\mu_0 H \tag{2}$$

The requirement for perpendicularity of the two fields can be explained as follows. The precessing magnetic moment can resonate only with a magnetic field with which it is in phase.

Consequently, if the alternating field is to change the orientation of the magnetic moment it must be rotated around the static field $\vec{H}$ at the precession frequency. In general the rotating field $\vec{H}_1$ can have components along $\vec{H}$ as well as components perpendicular to $\vec{H}$. The perturbing influence of the alternating magnetic field, however, is entirely determined by the perpendicular components. A linearly polarized field is usually used in practice instead of a rotating one; this can be thought of as the sum of two fields rotating at the same rate in two opposite directions.

The alternating field will cause transitions from lower levels to upper levels and vice versa with equal probability. Very frequently the population of the lower energy levels (number of particles with low energy) is greater than the population of the upper levels and transitions from lower to upper levels predominate, causing absorption of energy. The population of all levels finally evens out when the number of transitions from the lower to the upper levels becomes equal to the number of transitions in the reverse direction (we then have the so-called saturation condition). It follows from this that, in a system of isolated paramagnetic spins, steady-state resonant absorption does not exist.

However, if the assembly of magnetic moments of all paramagnetic spins (the spin system) can interact with the "lattice," which includes all other remaining degrees of freedom of the paramagnetic substance (interaction with magnetic systems of other types, interaction with diamagnetic systems, etc.), and can, as a result of such interaction, transmit the energy received from the RF field to the lattice, then the resonant magnetic absorption can be steady state and can be measured.

Let us examine the problem of interaction between the spin system and the lattice [9, 10]. We will assume initially that the RF field is absent and that the constant magnetic field is vanishingly small. In this case the amount of level splitting in the magnetic field is vanishingly small; i.e., the population of all the levels is about the same. If we now instantaneously increase the magnetic field, the levels will become separated. An equal population of the levels will correspond to a high energy of the spin system, and the temperature of the spin system will be much higher than the lattice temperature. The question arises of how fast thermal equilibrium is established between the spin system and the lattice; i.e., how much time does it take until the level population returns to the steady-state distribution. (Together with this redistribution will be established a magnetization of the material in the direction of the external magnetic field, the longitudinal magnetization.)

The problem can also be formulated in another way. Suppose that a constant field was applied and that thermal equilibrium has been established between the spin system and the lattice. The population of the lower levels is, of course, larger than that of the upper levels (there is magnetization in the direction of the magnetic field). Let us now apply a strong RF field at the resonant frequency, causing the population of the levels to become equal and bringing about a disappearance of the longitudinal magnetization. The question is, how long will the longitudinal relaxation last after the RF field is removed?

For the sake of simplicity we will consider only two states (i.e., $F = 1/2$). Let us denote the transition probability per unit time from the upper state to the lower state by $w_{21}$, that from the lower to the upper by $w_{12}$, the number of paramagnetic spins with higher energy (number of particles in the upper state) by $N_2$, and the number of spins in the lower state by $N_1$ ($N_1$ and $N_2$ are numbers of particles per cm$^3$). The condition for thermal equilibrium between the spin system and the lattice is that the number of transitions in both directions between the two states must be equal, i.e.,

$$N_2 w_{21} = N_1 w_{12} \tag{3}$$

At equilibrium the ratio of the populations $N_1$ and $N_2$ is given by the Boltzmann factor

$$N_1/N_2 = e^{\frac{\Delta E}{kT}} \tag{4}$$

where $\Delta E = 2\mu H$ is the energy spacing between the two states, and $kT$ is the thermal energy. At room temperature and field strengths of the order of 100 ka-turns/meter we have $2\mu H/kT \approx 10^{-4}$ to $10^{-6}$.* Expanding the exponential term in a series and considering only first-order terms, we get from Eq. (3)

$$w_{21} = w(1 + \mu H/kT)$$

$$w_{12} = w(1 - \mu H/kT)$$

where $w = \frac{1}{2}(w_{12} + w_{21})$ is the average value of the two transition probabilities.

Let us now return to the problem of longitudinal relaxation. At the initial moment the spin system and the lattice are not in equilibrium. The excess $n$ of particles per cm$^3$ in the lower

---

*In this book the international MKS system of units is used (All-Union State Standard 9867-61). In this system the unit of the magnetic field $H$ is the ampere per meter. This unit is related to the oersted (unit of $H$ in the Gaussian system) by 1 amp/m = $4\pi \times 10^{-3}$ oersted.

Translator's note: The unit ampere-turn/meter is used commonly in the USA for the magnetic field in the MKS (practical) system. This unit is the same as the author's ampere/meter and will also be used in the text.

state is $n = N_1 - N_2$. The rate of change of this excess per unit time is given by $+\dfrac{dn}{dt}$. On the other hand, this rate of change is equal to the difference between the number of transitions from the upper state to the lower and that from the lower to the upper multiplied by two (since for one transition the number of particles in one state decreases by one while it increases by one in the other state). Thus

$$\frac{dn}{dt} = 2(w_{21}N_2 - w_{12}N_1)$$

Substituting the values of $w_{21}$ and $w_{12}$ we get

$$\frac{dn}{dt} = 2w(n_0 - n) \tag{5}$$

where $n_0 = N\mu H/kT$, and $N = N_1 + N_2$ is the total number of particles per cm$^3$.

Integrating this equation we get

$$(n_0 - n) = (n_0 - n_a)e^{-2wt} \tag{6}$$

where $n_a$ is the initial value of $n$, and $n_0$ is the value of the excess at equilibrium (when $t \to \infty$).

The approach toward equilibrium of the spin system and the lattice (so called spin-lattice relaxation or longitudinal thermal relaxation) follows an exponential law with a time constant $\tau_1 = \frac{1}{2}w$. Consequently, the time $\tau_1$ is called the spin-lattice relaxation time or the thermal relaxation time.

When an oscillating magnetic field is present, the initial equation for $n$ becomes

$$\frac{dn}{dt} = \frac{n_0 - n}{\tau_1} - 2nW \tag{7}$$

where $2nW$ represents the difference between transitions from the lower state to the upper state and from the upper to the lower induced by the RF field; and $W$ is the probability per unit time of the induced transitions.

Integrating this equation we get

$$(n_s - n) = (n_s - n_a) \exp(-t/Z\tau_1) \tag{8}$$

Here $n_s = n_0 Z$, $Z^{-1} = 1 + 2W\tau_1$. When the radio frequency field is present the spin system approaches the steady state also according to an exponential law, but the time constant is now $\tau_1 Z$ instead of $\tau_1$. The quantity $n_s$ denotes the steady-state excess of particles in the lower state of the spin system $(t \to \infty)$.

Remembering that $n_0 = N\mu H/kT$ and similarity writing $n_s = N\mu H/kT_s$ we get

$$\frac{n_s}{n_0} = \frac{T_s}{T_0} = Z \quad \text{or} \quad T_s = T\frac{n_0}{n_s} = \frac{T}{Z}$$

Thus the equilibrium of the spin system in the presence of an applied radio frequency magnetic field does not correspond to thermal equilibrium between the spin system and the lattice. When an RF magnetic field is present, even at steady state the equilibrium temperature $T_s$ of the spin system is higher than the lattice temperature. The RF magnetic field seems to continuously "warm up" the spin system.

The probability of induced transitions has a frequency dependence $g(\nu)$ and is proportional to the intensity $H_1^2$ of the radio frequency magnetic field. Up to a factor $A$ we can write $W = Ag(\nu)H_1^2$. Then

$$Z^{-1} = 1 + 2A\tau_1 g(\nu)H_1^2$$

and

$$n_s = n_0 \left[1 + 2A\tau_1 g(\nu)H_1^2\right]^{-1}$$

If the amplitude of $H_1$ is large, the ratio $n_s/n_0$ is very small, corresponding to a high spin temperature $T_s$. In this case the spin system will be saturated ($n_s \to 0$; i.e., the population of both states is equalized). This is the reason why the quantity $Z$ is called the saturation factor. At constant values of $\nu$ and $H_1$, saturation occurs sooner for long spin-lattice relaxation times and later for short times.

The spin temperature can reach very high values—tens and hundreds of millions of degrees. Thus, for example, observation of nuclear resonance of the protons in ice at 88°K placed in a constant magnetic field of 560 ka-turns/m and a radio frequency field of 8 amp-turns/m yields a value of the saturation factor of the order of $10^{-6}$. This means that $T_s$ of the nuclear spin system in steady state will be of the order of $10^8$ °K. What energy is then required to obtain such a spin-system temperature? We will assume that complete saturation takes place and that the excess $n_s$ is zero. For each cubic centimeter we must supply energy equal to $n_0 \mu_p H$, where $\mu_p$ is the magnetic moment of the proton. But $n_0 = N \mu_p H/kT$, so that the required energy is $N(\mu_p H)^2/kT$. Calculations for the protons of ice at 88°K and $H = 560$ ka-turns/m show the required energy to be of the order of $10^{-7}$ joules/mole. This result points out that the specific heat of the spin system is extremely low compared with the specific heat of the lattice. In reality the temperature of matter will be determined by the temperature of the lattice.

The above mechanism of spin-lattice relaxation, characterized by the longitudinal relaxation time $\tau_1$, determines the establishment of the equilibrium value of magnetization along the direction of the external magnetic field; i.e., the time $\tau_1$ characterizes the appearance of a magnetic moment $\vec{M}$, different from zero, in the direction of the magnetic field (Fig. 5). If all spins were located at one point, their magnetic moments would precess together, without changing their relative distribution. In this case we would have a total moment $\vec{M}$ of all spins whose direction did not coincide with that of the external magnetic

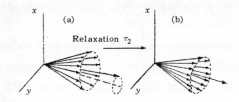

FIG. 5. Precession of magnetic moments of systems. a—Initial time; b—after completion of transverse relaxation

field (as shown in Fig. 5a, the magnetic moments are irregularly distributed on a cone) and the component of the moment $\vec{M}_\perp$, rotating in the $xy$ plane and representing the magnetization in the plane perpendicular to the constant magnetic field. The paramagnetic spins, however, have an extended spatial distribution and due to a number of reasons are located in somewhat different magnetic fields. Due to this the precession frequencies of the magnetic moments are different from one another and their phase coherence is not maintained. The direction of the total magnetic moment in this case approaches the direction of the constant magnetic field, and its projection on the $xy$ plane (which describes transverse magnetization) approaches zero. This process is called transverse relaxation and is characterized by the time $\tau_2$.

There are several factors contributing to the disturbance of phase coherence. Thus, due to nonuniform surroundings (different neighbors, nonidentical diamagnetic shielding, etc.), the paramagnetic spins are located in somewhat differing magnetic fields. Because of this, their precession frequencies are not quite identical, which causes disturbance of the phase coherence. An inhomogeneous external magnetic field will have a similar effect.

The spin-exchange processes also play an important part in transverse relaxation. This can be explained as follows. Suppose we have two spins, $j$ and $k$. Each of these spins is in the field of the other and both fields oscillate at the proper Larmor frequency. These oscillating fields can cause a transition, for example, in spin $k$. There will be a mutual exchange of energy since spin $k$ will receive energy from spin $j$; in other words, we will have a spin-exchange process. It is obvious that processes of this type will influence the phase coherence of the totality of precessing moments, causing a decrease in $r_2$. (In the case of nonidentical paramagnetic spins the spin-exchange process is absent because their precession frequencies are different.)

## 4. BLOCH EQUATIONS

Magnetic-resonance absorption can be described using the imaginary part of the complex magnetic susceptibility $\chi = \chi' - i\chi''$. The first fundamental work in this direction was done by L. D. Landau and Ye. M. Lifshits, who obtained $\chi$ as a function of frequency and described the behavior of ferromagnetic crystals in an oscillating magnetic field [5]. Further development of this subject is due to many researchers. In this section the behavior of a group of magnetic spins located in oscillating and constant magnetic fields will be investigated using the approach of Bloch [9, 10].

The magnetic moments of the individual spins in one cubic centimeter of substance make up the magnetization vector $\vec{M}$. Generally this vector will have components $M_x$, $M_y$ and $M_z$ and will be located in an external magnetic field whose components are

$$H_z = H_0, \quad H_x = 2H_1 \cos \omega t, \quad H_y = 0 \tag{9}$$

The RF field $H_x$ can be represented as the sum of two rotating fields. Of these two fields only the one whose direction of rotation coincides with the direction of precession of the magnetic moment $\vec{M}$ around the direction of the constant magnetic field $H$ can influence the magnetic moments, changing their orientation. Let us assume that this is the clockwise rotating field (in this case the choice is arbitrary), so that the components of the RF field are

$$H_x = H_1 \cos \omega t, \quad H_y = -H_1 \sin \omega t, \quad H_z = 0 \tag{10}$$

On the other hand, we know that a magnetic moment $\vec{M}$ placed in a magnetic field $\vec{H}$ will be acted upon by a torque $[\vec{M}\vec{H}]$. The time rate of change of vector $\vec{M}$ under the influence of this torque can be written in the form

$$\frac{d\vec{M}}{dt} = \gamma \left[\vec{M}\vec{H}\right]$$

where $\gamma$ is the ratio of the magnetic moment to the angular moment and is equal to $2\pi g \mu_0 / h$. Considering Eq. (10) we get

$$\frac{dM_x}{dt} = \gamma (M_y H_0 + M_z H_1 \sin \omega t)$$

$$\frac{dM_y}{dt} = -\gamma (M_x H_0 - M_z H_1 \cos \omega t) \tag{11}$$

$$\frac{dM_z}{dt} = -\gamma (M_x H_1 \sin \omega t + M_y H_1 \cos \omega t)$$

The system of equations (11) is, however, not sufficiently complete for the real spin system. The equations do not account for the processes of longitudinal and transverse relaxation. Due to longitudinal relaxation, $M_z$ increases to some equilibrium value $M_0$. The increase in $M_z$ is exponential with a time constant $\tau_1$. Therefore

$$\frac{dM_z}{dt} = \frac{M_0 - M_z}{\tau_1} \tag{12}$$

In a similar way, due to transverse relaxation the transverse components of magnetization approach zero with a time constant $\tau_2$. For simplicity it can be assumed that $M_x$ and $M_y$ approach zero exponentially, i.e.,

$$\frac{dM_x}{dt} = -\frac{M_x}{\tau_2} \quad \text{and} \quad \frac{dM_y}{dt} = -\frac{M_y}{\tau_2} \tag{13}$$

Thus the complete system of equations describing the behavior of $M_x$, $M_y$ and $M_z$ has the form

$$\frac{dM_x}{dt} = \gamma(M_y H_0 + M_z H_1 \sin \omega t) - \frac{M_x}{\tau_2}$$

$$\frac{dM_y}{dt} = -\gamma(M_x H_0 - M_z H_1 \cos \omega t) - \frac{M_y}{\tau_2} \tag{14}$$

$$\frac{dM_z}{dt} = -\gamma(M_x H_1 \sin \omega t + M_y H_1 \cos \omega t) + \frac{M_0 - M_z}{\tau_1}$$

Introducing new variables $u$ and $v$:

$$u = M_x \cos \omega t - M_y \sin \omega t, \ v = -(M_x \sin \omega t + M_y \cos \omega t) \tag{15}$$

so that

$$M_x = u \cos \omega t \ v \sin \omega t, \ M_y = -(u \sin \omega t + v \cos \omega t) \tag{16}$$

the system of equations (14) can be written as

$$\frac{du}{dt} + \frac{u}{\tau_2} + (\omega_0 - \omega)v = 0$$

$$\frac{dv}{dt} + \frac{v}{\tau_2} - (\omega_0 - \omega)u = -\gamma M_z H_1 \qquad (17)$$

$$\frac{dM_z}{dt} + \frac{M_z}{\tau_1} - \gamma H_1 v = \frac{M_0}{\tau_1}$$

where $\omega_0 = \gamma H_0$.

The steady-state condition is

$$\frac{du}{dt} = \frac{dv}{dt} = \frac{dM_z}{dt} = 0$$

Considering this, it is not difficult to solve system (17). Substituting the expressions obtained for $u$ and $v$ in (16) we get

$$M_x = \frac{\gamma \tau_2 H_1 M_0 [\tau_2(\omega_0 - \omega)\cos\omega t + \sin\omega t]}{1 + (\omega_0 - \omega)^2 \tau_2^2 + \gamma^2 H_1^2 \tau_1 \tau_2}$$

$$M_y = \frac{\gamma \tau_2 H_1 M_0 [\tau_2(\omega_0 - \omega)\sin\omega t - \cos\omega t]}{1 + (\omega_0 - \omega)^2 \tau_2^2 + \gamma^2 H_1^2 \tau_1 \tau_2} \qquad (18)$$

In a real experiment the oscillating field is not circularly but linearly polarized and is of the form $H_x = 2H_1 \cos\omega t$. Furthermore, it is known that an oscillating magnetic field $H_x = 2H_1 \cos\omega t$ gives an in-phase magnetization $2\chi' H_1 \cos\omega t$ and a quadrature component $2\chi'' H_1 \sin\omega t$. Taking this into account it is seen from Eqs. (18) that

$$\chi' = \frac{1}{2} \chi_0 \omega_0 \tau_2 \frac{(\omega_0 - \omega)\tau_2}{1 + (\omega_0 - \omega)^2 \tau_2^2 + \gamma^2 H_1^2 \tau_1 \tau_2}$$

$$\chi'' = \frac{1}{2} \chi_0 \omega_0 \tau_2 \frac{1}{1 + (\omega_0 - \omega)^2 \tau_2^2 + \gamma^2 H_1^2 \tau_1 \tau_2} \qquad (19)$$

where $\chi_0 \omega_0 \tau_2 = \gamma \tau_2 M_0$ since $M_0 = \chi_0 H_0$ and $\omega_0 = \gamma H_0$.

The static magnetic susceptibility can easily be expressed in terms of the magnitude of the magnetic moment, the temperature and the number of spins. In fact, taking for simplicity the case when the spins have $F = \frac{1}{2}$, we can write that the magnetic moment per cm$^3$ of substance is $(N_1 - N_2)\mu$, where $N_1$ is the number of spins in the lower state and $N_2$ is the number of spins in the upper state. But

$$\frac{N_1}{N_2} = e^{\frac{2\mu H}{kT_s}} \approx 1 + \frac{2\mu H}{kT_s} \tag{20}$$

Since the total number of spins is $N = N_1 + N_2$, we get

$$(N_1 - N_2)\mu \approx \frac{\mu^2 HN}{kT_s}$$

Therefore

$$M = \chi_0 H \simeq \frac{\mu^2 HN}{kT_s} \qquad \text{and} \qquad \chi_0 = \frac{N\mu^2}{kT_s}$$

Thus, finally, the components of the magnetic susceptibility are

$$\chi' = \frac{1}{2} \frac{N\mu^2}{kT_s} \omega_0 \tau_2 \frac{(\omega_0 - \omega)\tau_2}{1 + (\omega_0 - \omega)^2 \tau_2^2 + \gamma^2 H_1^2 \tau_1 \tau_2}$$

$$\chi'' = \frac{1}{2} \frac{N\mu^2}{kT_s} \omega_0 \tau_2 \frac{1}{1 + (\omega_0 - \omega)^2 \tau_2^2 + \gamma^2 H_1^2 \tau_1 \tau_2} \tag{21}$$

The absorbed power $P$ can be found as follows. The average rate of energy absorption per cm$^3$ of substance is equal to the average value of $H_x \dfrac{dM}{dt}$, where $H_x = 2H_1 \cos \omega t$, and $M$ is the magnetization (only the quadrature component of magnetization, $2\chi'' H_1 \sin \omega t$, contributes to absorption). Then

$$P = \frac{1}{T} \int_0^T H_x \frac{dM}{dt} dt = 2\omega \chi'' H_1^2$$

Thus in general

$$P = \frac{N\mu^2}{kT_s} \omega\omega_0 \frac{\tau_2 H_1^2}{1 + (\omega_0 - \omega)^2 \tau_2^2 + \gamma^2 H_1^2 \tau_1 \tau_2} \qquad (22)$$

On the other hand, the absorbed power can be written as a product of the difference in populations in the two states, $N_1 - N_2 \simeq N\mu H/kT_s$, the quantum of energy $\frac{h}{2\pi}\omega_0$, and the transition probability $W = Ag(\omega)H_1^2$ (see Section 3), i.e.,

$$P = Ag(\omega)H_1^2 h \omega_0 \frac{N\mu H}{2\pi kT_s}$$

or

$$P = A \frac{N\mu h}{2\pi \gamma kT_s} \omega_0^2 g(\omega)H_1^2$$

Comparing this expression with Eq. (22), in which we can let $\omega\omega_0 \approx \omega_0^2$, we conclude that the function $g(\omega)$ which describes the shape of the absorption line is

$$g(\omega) \approx \frac{\tau_2}{1 + (\omega_0 - \omega)^2 \tau_2^2 + \gamma^2 H_1^2 \tau_1 \tau_2} \qquad (23)$$

It is immediately evident from this relationship that the maximum of the absorption line and its half-width points are given by

$$g_{max} \simeq \frac{\tau_2}{1 + \gamma^2 H_1^2 \tau_1 \tau_2} \qquad (24)$$

$$(\omega - \omega_0)^2 = \frac{1}{\tau_2^2} + \gamma^2 H_1^2 \frac{\tau_1}{\tau_2} \qquad (25)$$

It follows from this that an increase in amplitude of the oscillating magnetic field $H_1$ causes a broadening of the absorption line and a decrease of signal (parameter $\gamma^2 H_1^2 \tau_1 \tau_2$ determines the degree of saturation). Consequently the absorption signal can be increased if $H_1$ is decreased. It is not possible to decrease $H_1$ without limit since, according to (22), the absorbed power is proportional to $H_1^2$. For constant $H_1$ the degree of saturation can be decreased by decreasing the spin-lattice relaxation time $\tau_1$. A decrease of the transverse relaxation time $\tau_2$ for this purpose is obviously not sensible since it will cause a broadening of the absorption signal. When saturation is absent, i.e., when $\gamma^2 H_1^2 \tau_1 \tau_2 \ll 1$, Eq. (25) defining the width of the line becomes

$$|\omega_0 - \omega| = \frac{1}{\tau_2}$$

Examining Eq. (22) and noting that the spin-system temperature is $T_s = TZ^{-1}$ (see Section 3), where $T$ is the temperature of the lattice, we conclude that an increase in temperature causes a decrease in absorption signal. This result is completely reasonable if it is noted that thermal motion prevents the orientation of the magnetic moment in the direction of the field.

*Chapter III*

# Electron Paramagnetic Resonance

## 5. THE PHENOMENON OF ELECTRON PARAMAGNETIC RESONANCE

Electron paramagnetic resonance (EPR) is one of the concrete forms of magnetic resonance discussed in the preceding chapter. The word "electron" emphasizes the fact that the resonant transitions occur between levels which depend on the interaction of the magnetic field and the magnetic moment of the electrons (electron shells of atoms or ions). Let us now explain the proper relationship of these levels to the general atomic energy level system. In quantum mechanics the atomic levels are determined in sequence by application of perturbation theory. In the first approximation we consider only the Coulomb interaction. Within this approximation the energy depends on $\vec{L}$ and $\vec{S}$, i.e., the orbital angular momentum and the total spin (assuming $LS$ coupling), and is independent of the magnetic quantum numbers $M_L$ and $M_S$. The energy levels of the atom (terms) can be written in the form $E = E_{LS}$. Because $M_L$ can have

$2L + 1$ values and $M_S$ can have $2S + 1$, the terms have a spatial degeneracy with multiplicity $(2S + 1)(2S + 1)$.

In the second approximation the interaction between the total spin angular momentum $\vec{S}$ and the total orbital angular momentum $\vec{L}$ is considered. Due to this interaction $\vec{L}$ and $\vec{S}$ are not conserved separately and only the total angular momentum $\vec{J} = \vec{L} + \vec{S}$ is conserved. Depending on the orientation of the total spin angular momentum of the atom with respect to the total orbital angular momentum, the levels are split and multiplets appear. In this case one speaks of the fine structure of the atomic energy levels. Thus $E = E_{LSJ} = E_{LS} + \Delta E_J$, where $\Delta E_J$ is the spacing between two components of a multiplet. When the spin-orbit interaction is considered the spatial degeneracy disappears partially. Actually, the energy levels now depend upon the mutual orientation of $\vec{L}$ and $\vec{S}$, i.e., $|\vec{J}|$, and are independent of the orientation of $\vec{J}$, so that each component of the multiplet is degenerate with multiplicity $2J + 1$.

In the third approximation the interaction between the nuclear magnetic moment and the internal atomic magnetic field is taken into account. It appears that in reality the fine-structure lines consist of a number of sublevels whose number is equal to the number of possible orientations of the magnetic moment of the nucleus with respect to the internal atomic field. The totality of such hypermultiplets constitutes the hyperfine structure of the atomic energy levels.

The superposition of an external magnetic field causes an orientation of the magnetic moment of the atom. To each such orientation there corresponds a specific energy level $E = g\mu_0 H M_J$; i.e., each level of the fine structure splits into $2J + 1$ sublevels (Section 3). Figure 6 shows a schematic

representation of the atomic energy levels in the presence of
an external magnetic field (the interaction with the nucleus is
not taken into account). It also shows possible transitions
between the levels of the fine structure. These transitions
should satisfy the well-defined selection rules which follow
from the laws of conservation of energy and angular momentum

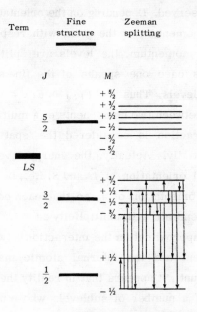

FIG. 6. Energy levels of an atom in the
presence of an external magnetic field
and possible transitions between the fine-
structure levels

and from symmetry considerations. The probability of radiation
(or absorption) of photons is different for photons with various
properties. The properties of photons can be characterized by
the energy $h\nu$, the spin $i = 1$, the orbital angular momentum with
respect to the atom $\vec{l}$, the total angular momentum $\vec{j}$, the projection of the total angular momentum and the symmetry. In

particular, for dipole radiation we have $j = 1$ and for quadrupole radiation we have $j = 2$. Presently we are interested only in magnetic dipole transitions. If before the transition the angular momentum of the atom is equal to $\vec{J}$ and the angular momentum of the quantum is $\vec{1}$, after the transition the angular momentum of the atom is $\vec{J'} = \vec{J} + \vec{1}$. Since addition of vectors $\vec{J}$ and $\vec{1}$ can be accomplished in various ways, the magnitude of the total angular momentum $\vec{J'}$ can have the values $J - 1, J$ and $J + 1$. Therefore, for a dipole transition the change in the total angular momentum of the system should satisfy the condition

$$\Delta J = 0, \pm 1 \qquad (26)$$

It should also be remembered that transitions between the states for which $J = 0$ and $J' = 0$ are impossible since $\vec{0} + \vec{1} \neq \vec{0}$.

The selection rules for the projection of the total moment are completely analogous:

$$\Delta M_J = 0, \pm 1 \qquad (27)$$

These rules follow from the fact that the change in the projection of the total angular momentum is due to addition (or subtraction) of the projection of the angular momentum of a photon $\vec{1}$. The projection of an arbitrary angular momentum $\vec{A}$ can have $2A + 1$ values: $A, A - 1, \ldots, -(A - 1), -A$, and therefore the projections of the angular momentum $\vec{1}$ are $1, 0, -1$ and the change in the projection of the total angular momentum is equal to $0, \pm 1$.

Using the obtained selection rules (26) and (27), we investigate schematically the possible transitions between two fine-structure levels (Fig. 6), assuming at the same time that they are split into a number of sublevels by a magnetic field. The first group of transitions satisfies the selection rules $\Delta J = \pm 1$, $\Delta M_J = 0, \pm 1$. The quantum of energy which is emitted or absorbed during these transitions may be sufficiently large that these transitions can be observed with an ordinary optical spectrometer. The appearance of transitions of this group was essential to the Zeeman effect in its old interpretation: without the field there was only one spectral line, and when the field was applied this line was split up into a number of finer lines.

The second group of transitions, which satisfies the selection rules $\Delta J = 0$, $\Delta M_J = \pm 1$, is of much greater interest to us. These transitions take place between the Zeeman components which belong to a group of levels with identical values of $J$, and they take place only within the limits of this one group. Within the limits of one groups the $g$ factor is constant and the Zeeman levels are equidistant. For every group of Zeeman sublevels, characterized by its value of $J$, the $g$ factor is different. Therefore the distance between the Zeeman levels in various groups will be different. Using the frequency as a measure of this distance we get

$$\nu = \frac{g \mu_0 H}{h} \simeq 3.10 \cdot 10^4 H \tag{28}$$

where $H$ is expressed in amp-turns/m and $\nu$ is in c/s. When $H = 10$ ka-turns/m we get $\nu \simeq 300$ Mc/s and $\lambda \simeq 100$ cm, when $H = 100$ ka-turns/m we get $\nu \simeq 3000$ Mc/s and $\lambda \simeq 10$ cm, and when $H = 1000$ ka-turns/m we get $\nu \simeq 30,000$ Mc/s and $\lambda \simeq 1$ cm. Therefore the transitions of the second group, or between the Zeeman components, fall within the microwave and radio frequency portion of the electromagnetic spectrum. These transitions can be observed by radio frequency methods, and it is with these transitions that we will deal in our study of electron paramagnetic resonance.

It is important to emphasize the role of the oscillating magnetic field in observations of EPR. Generally the transitions between the levels take place even without the application of an external oscillating field. Frequently the intensity of spontaneous transitions is sufficient for observation of the Zeeman effect in the optical region. With a decrease in frequency, however, the probability of such transitions is drastically reduced. For dipole transitions, for example, this probability is proportional to $\nu^3$. If in the optical region (transitions $\Delta J = \pm 1$) $\nu \simeq 10^{15}$ c/s then for transitions between the Zeeman components $(\Delta J = 0)$, $\nu \simeq 10^7$-$10^{10}$ c/s, and the probability of transitions is decreased by many orders of magnitude. At the same time the intensity of spontaneous radiation (or absorption) decreases by about the same amount. Observation of such negligibly low intensities becomes practically impossible. The absorption intensity can be increased, however, in an artificial manner if the atoms of the paramagnetic substance are excited by an alternating magnetic field of resonant frequency which is perpendicular to the static magnetic field. During such excitation, absorption of quanta of electromagnetic energy will take place. It is also necessary to assure that the alternating magnetic field will be relatively weak in order to prevent saturation (Chapter 2). Furthermore, EPR can be observed only in macroscopic systems in which spin-lattice relaxation takes place.

## 6. METHODS OF OBSERVING EPR

The present methods for investigating EPR are based on observation of some parameter of a radio circuit whose change is related to changes in the RF field, which in turn are caused by EPR. The theory of the method establishes a connection between the changes in the radio circuit parameters and the quantities which characterize the electron paramagnetic resonant absorption (most often absorption $\chi''$ is considered). The theory of EPR then connects these macroscopic quantities with the microscopic constants of matter (spin, magnetic moments, etc.).

From the experimental viewpoint all methods of observing EPR can be divided into two large groups: 1) methods operating in the microwave region ($10^{10}$-$10^{11}$ c/s) and 2) methods operating in the radio frequency region ($10^6$-$10^9$ c/s). The differences between these two groups are determined by differences in the circuits utilized, the transmission lines, the methods of generating the high-frequency field, etc. The methods belonging to the first group are more popular because they are generally much more sensitive than the methods of second group.

With respect to sensitivity all radio spectrometers can be conditionally divided into three groups: 1) high-sensitivity radio spectrometers, with which one can detect less than $10^{-9}$ moles of diphenylpicrylhydrazyl (DPPH), 2) intermediate-sensitivity radio spectrometers, with which one can detect $10^{-7}$-$10^{-9}$ moles of DPPH, 3) low-sensitivity radio spectrometers, with which one can detect not less than $10^{-7}$ moles of DPPH. Di-

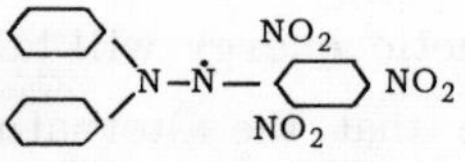

FIG. 7. Structural formula of diphenylpicrylhydrazyl

phenylpicrylhydrazyl (structural formula in Fig. 7) is a generally accepted calibration standard for measurement of EPR spectra. One milligram of DPPH contains approximately $1.5 \times 10^8$ uncoupled electrons.

*Simple microwave radio spectrometers.* For studies of EPR absorption one can first use the usual methods for measurement of the magnetic properties of matter, in particular, the method of standing waves [11]. The essence of this method is that introduction of a sample into a waveguide changes the standing-wave pattern, whereby the minima of the field are displaced and the standing-wave ratio changes. From changes in these parameters it is possible to determine the absolute values of real and imaginary parts of the magnetic susceptibilities $\chi'$ and $\chi''$. The main disadvantage of this method is its low sensitivity, its complexity (the absorption line is measured point by point) and the difficulty of its application to studies of narrow EPR lines. Radio spectrometers of this type are not frequently used.

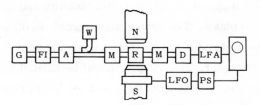

FIG. 8. Simple resonant-cavity radio spectrometer

Let us now investigate the most typical and presently the most widely used radio spectrometer. Figure 8 shows a block diagram of a simple spectrometer with a resonant cavity [12-14, 28]. The principle of operation is as follows. The microwave power, after going through resonator R, reaches detector D; after detection the signal is amplified by the low-frequency amplifier LFA and applied to the vertical deflection plates of an oscilloscope. When the intensity of the magnetic field changes from its value at resonance due to paramagnetic absorption, the circuit $Q$ decreases and the cavity and the waveguides are mismatched, causing the detector power input to decrease. In the figure G is the microwave generally (usually a klystron), which operates at a constant frequency and uses a stabilized power supply. FI is a ferrite isolator, which has the valuable characteristic of transmitting energy practically only in one direction. Use of microwave isolators greatly enhances the amplitude and frequency stability of the microwave power. In addition A is the attenuator, W is the wavemeter, and M are the matching networks. The meaning and principles of operation of these parts are clear and will not be explained. It is necessary to consider in greater detail the resonant cavity R in which the investigated sample is placed. It is desirable that the cavity be small in size and that it have high $Q$. With small cavities the required magnetic field intensity, stability and homogeneity are easier to obtain. The second requirement is dictated by the fact that the higher the cavity $Q$ the higher the sensitivity of the radio spectrometer. The sample should be located at the maximum of the cavity magnetic field and as far as possible from electric field maxima (since dielectric losses can significantly lower the cavity $Q$). With cylindrical cavities, the $H_{111}$* propagation mode is usually used. Rectangular cavities are also widely used, with the $H_{01n}$ propagation mode as a rule.

The sweeping of the magnetic field is accomplished with the help of the low-frequency oscillator LFO. In order to obtain sufficient modulation depth of the magnetic field (which is required, for example, in case of wide lines or in the case when the exact value of the magnetic field at resonance is unknown), low modulation frequencies are used since the inductance of the modulation coils is fairly large due to the influence of the poles of the magnet. Frequencies of the order of 50 c/s are usually

* $TE_{111}$ in U.S. terminology—Translation Editor.

used. (The minimum modulation frequency is determined by the persistence time of the oscilloscope screen.) When the magnetic field passes through the resonance value (or, as is said, when it passes the absorption line), paramagnetic absorption of microwave energy will take place in the cavity. In other words, the high-frequency signal is modulated and the microwave absorption signal is formed, after which the microwave power is fed to the detector. The form of the detected signal will correspond to the absorption line and will be repeated twice during each modulation period. (Passing the absorption line by changing the magnetic field is preferable to that by changing the microwave oscillator frequency since this eliminates the difficulties associated with wide-band oscillators and the dependence of the output power on the frequency; it also makes tuning of the waveguide much easier.) The detected microwave signal is fed to a low-frequency amplifier LFA. The bandwidth of the LFA should be sufficiently large to pass all frequency components (Fourier series) of the signal. For example, if the amplitude of the modulation function of the magnetic field is of the order of 15 ka-turns/m and if the absorption line width is $\sim 0.5$ ka-turn/m then the amplifier bandwidth required to amplify the signal with small distortion is about 10 kc/s. From the LFA the signal is fed to the vertical deflection plates of the oscilloscope whose sweep signal is taken from the magnetic field modulation voltage generator. For superposition of the direct and reverse variation this voltage is fed through the phase shifter PS.

The main advantage of the above radio spectrometer is its simplicity and the possibility of visual observation of the absorption lines on the screen of the oscilloscope. Such a spectrometer can be used for calibration and measurements, for conducting a quick search for new absorption lines, and for all investigations where high sensitivity is not required. The main disadvantage of this type of spectrometer is its low sensitivity, which is due to the large low-frequency noise level and low signal-to-noise ratio due to the wide bandwidth of the low-frequency amplifier.

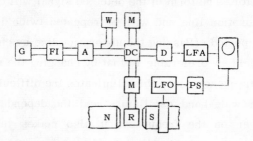

FIG. 9. Simple radio spectrometer with directional coupler

The second typical example of a simple ratio spectrometer is a spectrometer with a directional coupler (Fig. 9) [13-15, 28]. The directional coupler usually takes the form of a "magic T" or a "rat race" (Fig. 10). The characteristic property of

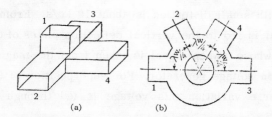

FIG. 10. Directional couplers. a—Magic T; b—rat race

these devices is that the power from the oscillator that is fed into the coupler appears in arms 2 and 3 but not in arm 4. A resonant cavity, placed between the poles of a magnet, is connected to arm 2 through a matching network. The power reflected from the resonant cavity goes into arms 1 and 4. The portion of the power which goes into arm 1 can be neglected (it is absorbed by the attenuator and the ferrite isolator). The portion of the energy entering arm 4 is compensated to a certain optimum level. The amplitude and phase compensation is accomplished by means of the matching network in arm 3 (usually an attenuator and a shorting plunger). At the moment when the resonant value of the magnetic field is reached, the magnitude of the power reflected from the resonant cavity changes and consequently the magnitude of the power entering arm 4 changes; i.e., the power delivered to the crystal detector changes. Detection results in a low-frequency pulse which, after amplification by the LFA, is fed to the vertical deflection plates of the oscilloscope. The sweep signal for the oscilloscope is obtained, as previously, from the generator of the magnetic field modulation function.

It is very important to know the residual amount of the unbalance of the bridge since the shape of the observed signal depends on this unbalance [16]. For clarity, let us assume that the unbalance exists only in amplitude and can be represented by the segment $QP$ of Fig. 11a ($OP$ corresponds to the signal from the arm containing the cavity and $OQ$ to the signal from the opposite arm). We will also assume that $QP \ll OP$. At resonance the tip of the vector $OP$ describes a small circle and the voltage fed to the amplifier changes with the magnetic field in a way similar to that shown in Fig. 12a. On the other hand, when the phases are unbalanced (Fig. 11b), the variation of the detector output voltage with the magnetic field has a shape similar to that shown in Fig. 12b. It should be noted that, due to balancing of the microwave power, the absorption signal is extracted in its pure form. This permits one to decrease the

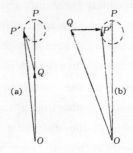

FIG. 11. Pure amplitude (a)
and pure phase (b) residual
unbalance [16]

general level of the power fed to the detector, thus decreasing the crystal detector noise, because the detector noise is directly proportional to the magnitude of the detected current. It should also be mentioned that with a decrease in power delivered to

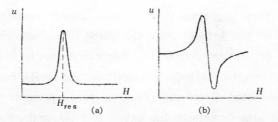

FIG. 12. Variation of the detector output voltage with the
magnetic field intensity in a simple radio spectrometer with
a directional coupler

the detector the conversion loss increases and, correspondingly, the sensitivity of the radio spectrometer decreases. The optimum conditions correspond to a crystal current of 0.5 ma, i.e., an input power level of the order of 1 mw (Fig. 13).

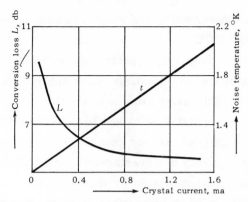

FIG. 13. Conversion losses and noise temperature of a
crystal as a function of detected current [12, 13]

*Methods for increasing the sensitivity of microwave radio
spectrometers.* The fundamental causes of low sensitivity in
radio spectrometers are insufficient stability of the absorption
signal and the high level of low-frequency noise. In order to
assure stability of the signal it is necessary that all components
of the radio spectrometer be highly stable and, primarily, that
the oscillator frequency and magnetic field frequency be very
stable.

In medium-sensitivity spectrometers the frequency stability
is usually maintained by electronically stabilizing the voltage
and stabilizing the heater supply current. Application of two-
stage electronic stabilizers (whose instability can be made as
low as $10^{-5}$-$10^{-6}$) and placing the klystron in a water-cooled oil
bath (to stabilize the temperature of the klystron) can lower the
frequency drift of the klystron in the 3-cm region to less than
1 Mc/s per hour, i.e., $\Delta f/f \sim 10^{-4}$. When the oil bath is not
used, the frequency drift can become greater than 2 Mc/s per
hour. The microwave oscillator frequency is usually measured
by precision wave meters ($Q \approx 20,000$), calibrated with a quartz

frequency standard, which are subject to measurement errors of the order of $10^{-4}$.

In high-sensitivity radio spectrometers, which are used for especially accurate measurements, more sophisticated methods of frequency stabilization are used. Usually the frequency of the microwave oscillator is locked to either the resonant frequency of a high-$Q$ cavity (for example, Pound's method) or to the standard frequency of a quartz oscillator [13]. The frequency measurement is accomplished by the use of special equipment and is usually based on comparison of the oscillator frequency with a known standard. The error in determination of the oscillator frequency in this case may be as low as $10^{-6}$-$10^{-7}$.

The above requirements also apply to magnets, which are the most expensive part of a microwave spectrometer for observing EPR. Stabilization of the magnetic field is achieved by stabilizing the electromagnet current with various types of current stabilizers. The instability of the magnetic field in this case is of the order of $10^{-4}$. As a rule, the magnetic field is measured by a flux meter whose operation is based on proton resonance (Section 11).

In situations where an extremely accurate measurement is needed, current stabilization of the magnetic field is not sufficient. It becomes necessary to account for the effect of temperature on the magnetic field (changes in temperature of the magnet are due to changes in the ambient temperature as well as to self-heating). Therefore, it is more feasible to stabilize the magnetic field directly, using again the phenomenon of proton resonance, and discarding the current stabilization. The instability of the magnetic field in this case can be lowered to $10^{-6}$.

The most difficult problem is that of decreasing the internal noise of the spectrometer, with the detector being the main source of low-frequency noise. Silicon-tungsten crystal diodes are most often used as detectors. These crystal detectors have excess noise; the magnitude of the noise is inversely proportional to the frequency of the output voltage and increases with increase of the detector power input.

The internal detector noise can be decreased if bolometers are used instead of crystals. The main advantage of bolometers is that they have a much lower excess-noise level. Furthermore, up to a certain level the bolometer noise is independent of the power input. Bolometers also have great disadvantages: at small power levels they have large conversion losses and their optimum operating level is about $10^{-2}$ watts. This relatively high power level causes some definite difficulties: small changes in position of matching stubs or in phase shifters cause a large change in signal. Moreover, bolometers are very sensitive to overload and have a short life. Thus the use of bolometers in radio frequency spectroscopy is rather rare.

The most effective method for increasing the sensitivity of a radio frequency spectrometer is to decrease the level of low-frequency detector noise. This can be achieved by decreasing the receiver bandwidth or by eliminating the low-frequency amplification altogether. The bandwidth of the amplifier chain should be sufficient to pass the required frequencies obtained from the Fourier analysis of the line shape. The spread of these frequencies depends on the width of the absorption line and the frequency and amplitude of the modulating field. The amplitude of the modulating field determines the shape of the signal: at large modulation depth a complete absorption line will be observed, but if the modulation depth is smaller than the width of the line, only a portion of the line will be observed. The frequency of the modulating field determines the rate of passage through the line and its repetition frequency. Figure 14 shows the effect of the rate of change of the magnetic field on the shape

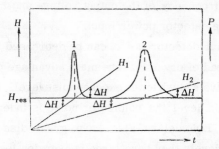

FIG. 14. Variation of the absorption signal shape
with the rate of change of the magnetic field

of the absorption signal. It is obvious that in the second case (curve 2) a much smaller number of Fourier harmonics is required for distortionless absorption signal amplification than in the first case (curve 1). Therefore, by decreasing the magnetic field modulation frequency, one can decrease the bandwidth of the amplifiers which follow the detector. If for a modulation frequency of 30-50 c/s (the time for the traversal of the narrowest lines in this case is about 30 $\mu$s) the required bandwidth is about 30-40 kc/s (and sometimes it is 1 Mc/s for better reproduction of the line shape) then for a slowly changing magnetic field the bandwidth can be decreased to 1 c/s or less. This results in a significant decrease in the level of low-frequency noise and, consequently, an increase in the sensitivity of the radio frequency spectrometer. It is obvious that in the case of a slowly changing magnetic field a graphc recorder should be used instead of an oscilloscope.

From the point of view of increasing the sensitivity of a radio frequency spectrometer, the application of phase-sensitive detection is very effective [9, 13]. The principle of operation of a phase-sensitive detector (PSD) is most conveniently explained with the help of the mechanical model shown in Fig. 15.

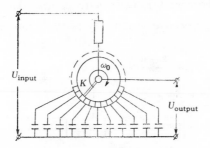

FIG. 15. Idealized diagram of a synchronous detector with capacitive integrator and mechanical commutation

The input signal is applied to the rotor, which rotates at the frequency corresponding to the speed of passage through the line. Every capacitor of the capacitance integrator is charged to some value which corresponds to some absorption signal amplitude. The low-frequency noise also appears at the input of the synchronous detector, but since the noise is a random function of time it will be averaged in the capacitive integrator. Now if the voltage input to the capacitive integrator is disconnected we get a signal which is free of random noise. In reality, a PSD is an electronic circuit (essentially a mixer) whose input consists of the signal voltage and the magnetic field modulator output.

Figure 16 is a block diagram of a system which can be used to observe the absorption signal in the case of low-frequency modulation of the magnetic field and phase-sensitive detection. This system can be used with any of the above-mentioned simple radio frequency spectrometers. From the detector the signal is fed to a narrow-band LFA and then to the PSD. From the output of the PSD the signal goes to the recorder. Application of this system significantly increases the sensitivity of radio frequency spectrometers. In particular, we have described a radio frequency spectrometer (with a transmission resonant cavity and a bolometer) whose sensitivity is $10^{-11}$ moles of DPPH [18].

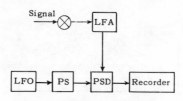

FIG. 16. System for observing EPR signals in the case of low-frequency modulation and phase-sensitive detection

Even greater advantages are available from the method of double modulation of the magnetic field, eliminating the low-frequency amplification [13, 14]. The method is based on the superposition of high-frequency modulation, whose amplitude is smaller than the width of the absorption line, on the slowly varying field of the electromagnet which covers the entire line. Figure 17a shows the variation of the detector current for the case of double magnetic field modulation (when a transmission resonant cavity is used). Further signal amplification takes place at the modulation frequency. The amplitude of the signal at any instant of time depends on the intensity of the slowly varying magnetic field at that instant (Fig. 17b); i.e., it is proportional to the slope of the line. As a result of this, a signal is produced whose envelope is proportional to the first derivative of the absorption line. In cases where it is desirable

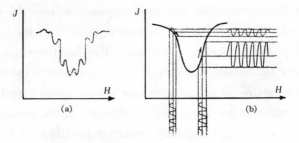

FIG. 17. Double modulation of the magnetic field

to obtain a line with minimum distortion, the modulation amplitude should not exceed 1/10 of the width of the observed line. Unfortunately, a decrease in the modulation amplitude causes a decrease in the amplitude of the output signal, i.e., a decrease in the sensitivity of the spectrometer. Therefore, in a number of important cases a compromise must be made: the modulation amplitude is increased (maximum signal is obtained when the modulation amplitude is comparable to the width of the line), even though this causes some distortion in the shape of the line.

Figure 18 is a block diagram of the system for observing paramagnetic resonant absorption utilizing double modulation of the magnetic field. This scheme can be used in radio frequency spectrometers with a transmission resonant cavity as well as in spectrometers with a waveguide directional coupler. The low-frequency modulation of the magnetic field is obtained either by

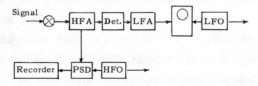

FIG. 18. System for observing EPR using double modulation of the magnetic field

using a large amplitude and low-frequency voltage (30-60 c/s) or by slowly sweeping the magnetic field with the help of a motor. In the first case the signal after amplification at high frequency and detection is amplified again and fed to the vertical deflection plates of an oscilloscope whose sweep signal is obtained from the low-frequency oscillator. In the second case, the signal after high-frequency amplification (HFA) is fed to the phase-sensitive detector and then to the recording chart, which moves synchronously with the magnetic field. The reference voltage for the PSD is obtained from the high-frequency magnetic field modulation function generator (HFO). The modulation frequency is usually chosen to be of the order of a few hundred kc/s. The achievement of the high-frequency modulation of the magnetic field is the main practical difficulty in work with radio frequency spectrometers of this type. Generation of a high-frequency field inside the cavity requires a powerful generator since at high frequencies the field does not penetrate the metallic walls of the cavity very well. Moreover, the "outside" high-frequency modulation results in a strong pickup of this frequency by all elements of the system. Therefore, the radio frequency field should be introduced inside the cavity either through a loop of thick wire or a partial slit in the resonant cavity.

The literature contains several descriptions of radio frequency spectrometers of this type. One of these [19] used a transmission resonant cavity excited in the $H_{011}$ mode (cavity diameter 45 mm, length 32 mm, $Q$ 8000, slit width 2.5 mm, coupling hole diameter 6 mm). The high-frequency modulation was at 975 kc/s, with a modulation amplitude of $\sim 0.15$ ka/m (with a current of the order of 40 ma). The low-frequency modulation was at 50 c/s, with a modulation depth of about 25 ka/m. Automatic frequency control was used to stabilize the klystron frequency with respect to the resonant frequency of the cavity. The sensitivity of the described radio frequency spectrometer was of the order of $4 \times 10^{-10}$ moles of DPPH when an oscilloscope was used and about $8 \times 10^{-12}$ moles when a graphic recorder was used.

Finally, let us consider one more method of increasing sensitivity which is very popular—the method of superheterodyne detection [13, 14]. Superheterodyne radio frequency spectrometers can have either a resonant cavity or a directional coupler. Their main advantage is that the signal amplification can take place at a sufficiently high intermediate frequency. Due to this the noise level is drastically reduced even at significant values of the bandwidth. All advantages of the superheterodyne detection method are best exhibited in radio frequency spectrometers with a directional coupler. This is because for the normal performance of a superheterodyne receiver it is required that the signal power be much smaller than the power delivered to the mixer from the local oscillator. Therefore, when the transmission resonant cavity is used, the power input to the cavity must be attenuated, causing a definite decrease in sensitivity. In the case of a directional coupler this power attenuation is not necessary. Figure 19 is a block diagram of the recording portion of the superheterodyne radio frequency spectrometer. From the crystal detector the intermediate-frequency signal (usually 30-45 Mc/s) passes through an intermediate frequency amplifier (IFA) and after detection and amplification is fed either to an oscilloscope or through a phase-sensitive detector to a recording chart.

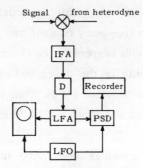

FIG. 19. Superheterodyne detection
system

The choice of intermediate frequency in superheterodyne radio frequency spectrometers is based on the following considerations. For a given amplitude of the rectified current the excess noise generated by the crystal is inversely proportional to the frequency (Fig. 20) (curve 1); i.e., as high as possible an intermediate frequency is desirable. On the other hand, the noise figure of intermediate-frequency amplifiers increases with frequency (curve 2); i.e., the intermediate frequency should not be too high. The minimum of the total noise figure is in the region 30-45 Mc/s. The sensitivity of superheterodyne radio frequency spectrometers reaches $10^{-10}$ moles of DPPH and higher.

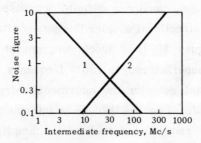

FIG. 20. Total noise figure due to noise generated in the crystal and the intermediate-frequency amplifier

*Methods of observing EPR in the radio frequency region.*
The most widespread, simple and also sensitive radio frequency
spectrometer for observing EPR in the radio frequency region
is the first radio frequency spectrometer, conceived by Ye. K.
Zavoyskiy [20]. This method of observing EPR is sometimes
also called the method of oscillator reaction [20-22].

The basic and almost the only element of this device is the
familiar vacuum-tube high-frequency power oscillator (Fig. 21).
The sample is placed in the circuit coil, which in turn is located
in the magnetic field. The magnetic field should be perpendicular

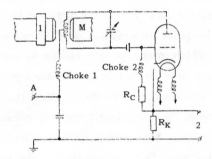

FIG. 21. Circuit of a radio frequency spectrometer for observation of EPR in the region of
$10^6$-$10^9$ c/s. 1—Modulation winding of the magnet,
2—output to a low-frequency amplifier or a
galvanometer [21]

to the magnetic component of the high-frequency magnetic field.
When the field of the magnet is varied, the paramagnetic absorption changes and, when certain conditions are satisfied, the
oscillator power load changes. This causes a change in the
grid and anode currents that is proportional to the absorption.
If the load change is small compared with the oscillator power,
the variations in the grid and anode currents and in the load are
linearly related. Since the paramagnetic losses are of the

order of several millionths of a watt, a linear relationship is always observed in practice. The linearity of the system can be checked by measuring the intensity of the resonant line in specimens which contain different quantities of paramagnetic substance mixed with a diamagnetic material in such a way that the volume of all samples is identical.

The variations in grid current are amplified by a DC amplifier and fed to a galvanometer. The absorption line in this case is measured point by point. In order to extract the grid current variations due to paramagnetic absorption, the actual grid current is balanced by a special bridge circuit. The sensitivity of the galvanometer should be of the order of $10^{-9}$.

The point-by-point observation of EPR has many disadvantages due to which many people refuse to use it and instead use the method of observation of the absorption lines on the screen of an oscilloscope. In this method the magnetic field modulation should be sufficiently deep to completely overlap the absorption line. The signal in this case is amplified by a low-frequency amplifier and applied to the vertical deflection plates of an oscilloscope. The sweep of the oscilloscope is obtained from the voltage modulating the magnetic field of the magnet.

The radio frequency spectrometer discussed above is exceedingly simple and readily available. In its application, however, it is necessary to account for the influence of external fields and to minimize this influence if possible. In particular, the oscillator must be shielded, all supply leads must contain RF chokes and be dressed against the shield, and the shield must be grounded. Furthermore, for precise measurements the entire apparatus—oscillator, amplifier and power supply (most frequently storage batteries)—is sometimes placed in a thick iron shield for additional protection from external magnetic and electric fields. The sensitivity of a radio frequency spectrometer utilizing the oscillator reaction method reaches quite high values, not lower than $3 \times 10^{-9}$ watts.

The observation of EPR can be accomplished through investigation not only of the absorption $\chi''$ but also of the dispersion of magnetic susceptibility $\chi'$. One of the first radio frequency spectrometers of this type was designed by Ye. K. Zavoyskiy [23]. The central part of this radio frequency spectrometer consists of two crossed coils, which are almost at right angles, placed in a constant magnetic field. Figure 22 explains the idea behind this method. Here $AB$ and $CD$ are the axes of the two coils and the angle $a$ between them is equal to or very close to $\pi/2$. Coil $AB$ is the transmitter of high-frequency power with a current of frequency $\nu$ flowing through it and generating a high-frequency magnetic field whose direction is along the axis of the coil. The component $H_{\sim\perp}$ of the

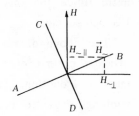

FIG. 22. Zavoyskiy's method of two crossed coils

high-frequency magnetic field that is perpendicular to the constant magnetic field $H$ will induce transitions between the Zeeman-split levels; i.e., it will reorient the magnetic dipoles. In the second, or receiving, coil $CD$ at the same time will be induced an emf which is proportional to the number of "elementary magnets" reoriented per unit time, i.e., proportional to the absorption. Under nonresonance conditions the transmitting and receiving coils are almost isolated and the emf induced in the receiving coil is very small. At resonance this emf increases significantly and can be amplified and recorded. For this purpose an instrument for measurement of the induced emf (e.g., a detector with a galvanometer) is connected to the coil *CD*. In this way the dispersion of the magnetic susceptibility can be studied. For some specific conditions the discussed method of two crossed coils can also be applied to the observation of absorption. The conditions depend on the coupling between the coils and their location in the external field.

The main disadvantage of the above method is the unfortunate position of the coils. The constant magnetic field intersects the turns of the receiving coil, with the result that the slightest changes in this field affect the results of measurements. Thus it is quite impossible to use the methods with oscillating magnetic fields for the observation of EPR signals on an oscilloscope screen.

The above disadvantages are absent in a method of two crossed coils proposed by Bloch [9] and called the method of nuclear induction. (The method was proposed for studies of nuclear magnetic resonance but is also applicable to studies of EPR.) In Bloch's method the axes of the coils and of the constant magnetic field are located along three mutually perpendicular directions. Due to this the changes in magnetic field cannot induce an emf in the receiving coil. Thus the magnetic field can be swept through resonance and the curve observed on the screen of an oscilloscope. Both the crossed coils and the sample container are placed in a special device called a probe for observation of magnetic resonance. The probe is also equipped with a current control to regulate the coupling current between the two coils, enabling the extraction of the real or imaginary part of the magnetic susceptibility, as the need arises. The block diagram of the system for observation of magnetic resonance by Bloch's method is shown in Fig. 23. The specimen is located inside the receiving coil and its dimensions are chosen in such a way as to obtain an optimum filling factor. The transmitter coil, whose axis is perpendicular to the axis of the receiver coil and to the direction of the magnetic field, is then wound on the outside. The receiver coil is tuned by means of a condenser, both of which make up the input tuned circuit of the radio frequency amplifier. The signals induced in the receiving coil by the precession of the resultant magnetization vector are usually very small but can be amplified before detection so that the obtainable signal-to-noise ratios are high. After detection and low-frequency amplification the signals can be fed to the input of an oscilloscope or, in the case of weak signals, to the input of a synchronous detector and a recorder. A low-frequency oscillator is used to modulate the constant magnetic field and synchronize the oscilloscope sweep, as well as for a reference signal for the synchronous detector.

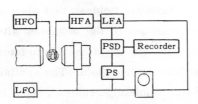

FIG. 23. Block diagram of Bloch's system [9]

Besides the Bloch method, several other methods, which are usually used in investigations of nuclear magnetic resonance, can be applied to studies of electron paramagnetic resonance in the radio frequency region (Section 11).

## 7. EFFECT OF THE INTERNAL CRYSTAL FIELD

Let us now examine the electrostatic interaction between the ions and its effect on EPR lines. Every ion is located in the electrostatic field of the surrounding ions. Due to shielding, the effect of this internal crystal field is always weaker than the Coulomb interaction between the electrons in an atom, but it appears in the interaction with a nonspherical electron cloud (i.e., in the case when the paramagnetic ion is not in an $S$ state). The problem of the effect of an external electrostatic field on the spectral lines of an atom is obviously a statement of the Stark effect. In our case the external electrostatic field is the crystal field. The splitting of levels caused by the crystal field can be computed with the help of perturbation theory. The perturbation operator is

$$H = -\vec{d}\vec{E}_{cr}$$

where $d$ is the dipole moment operator of the system.

The Stark effect is in many respects analogous to the Zeeman effect. The orientation of the magnetic moment is arbitrary in the absence of an external field, so that when an external field is applied the orientations of the moments with respect to the direction of this field are quantized due to interaction between the electron and the electric field. The spatial degeneracy is either partially or completely removed, depending on the magnitude and symmetry of the applied field and on the spin of the paramagnetic atoms.

Three separate cases are usually distinguished [13]. In the first case the crystal field is relatively weak, the spin-orbit coupling is not broken, and the total angular momentum is conserved. Now, is the spatial degeneracy completely removed by the crystal field and is observation of paramagnetic absorption possible? To answer this question we will start with two general theorems which we state without proof [13]. According to the first theorem, any arbitrary system with a degenerate ground state is spontaneously subjected to perturbations such that the system will achieve a minimum possible degree of degeneracy (Jahn-Teller effect). The second theorem states that in a system with an odd number of electrons located in an external electric field, the levels will always have even degeneracy (Kramers theorem).

It follows from these theorems that if the total angular momentum $J$ is a half-integer (odd number of electrons) then at least the lowest level is doubly degenerate; i.e., in this case paramagnetic resonant absorption is possible when the substance is placed in an external magnetic field. If the total angular momentum $J$ assumes integral values (even number of electrons), the multiplicity of the ground state is equal to unity (i.e., the state is nondegenerate) and the observation of paramagnetic resonant absorption is generally impossible. Due to the weakness of the internal crystal field, however, the splitting caused by this field is small. If this splitting is sufficiently small, paramagnetic resonance is possible even when $J$ is an integer. In this case the resonance levels will behave as doublets separated by a distance determined by the magnitude of the splitting connected with the internal crystal field. Such situations are observed in rare earth elements, where an incomplete shell with nonzero magnetic moment is located inside and partially shielded from the effect of the internal crystal electric field. The second case occurs when the field of the crystal is sufficiently strong to break the spin-orbit coupling. In this case

the angular momenta $L$ and $S$ are independent and, because of
the interaction of the orbital angular momentum with the crystal
field, the orbital degeneracy (of multiplicity $2L + 1$) is completely removed. The magnitude of the splitting is large (of the
order of $10^3$ cm$^{-1}$) and in practice at medium temperatures
only the lowest level is populated. This phenomenon is known
as "quenching of orbital levels." In particular it occurs in
elements of the iron group. The lowest level will have a spin
degeneracy of multiplicity $2S + 1$. According to the Kramers
and Jahn-Teller theorems the spin degeneracy is also removed:
in the case of an even number of electrons it is removed completely, while for an odd number of electrons double degeneracy
is preserved. Because of this EPR is only possible when the
number of electrons is odd.

The third case occurs when the internal crystal field is such
that it breaks the coupling not only between the angular momenta
$L$ and $S$ but also between the orbital angular momenta and spins
of individual electrons. In this case the orbital motion is
quenched and the spins tend to form pairs according to the
Pauli principle. It is obvious that in this case EPR is only
possible when the number of electrons is odd.

Thus the possibility of detection of electron paramagnetic
resonance and the observed results depend on the basic spectroscopic state of the ion, the magnitude of the spin-orbit
coupling, and especially the nature of the crystal field.


## 8. WIDTH OF EPR LINES

The width of the line is a very important factor which determines the possibility of observation of EPR. Very wide lines

are difficult to record and, furthermore, a large line width prevents resolution of fine and hyperfine structure.

In the first place, the width of the spectral line depends on the lifetime of the atom in its excited state. This follows directly from the energy uncertainty principle

$$\Delta E \cdot \Delta t \sim \frac{h}{2\pi} \tag{29}$$

This relationship expresses the fact that the law of conservation of energy can be verified by two measurements only with an accuracy of the order of $h/2\pi\Delta t$, where $\Delta t$ is the interval between the two measurements. (This should not be confused with the uncertainty relationship $\Delta p \cdot \Delta x \sim \frac{h}{2\pi}$; $E$ and $t$ have definite values at the identical moment.) In our case $\Delta t$ is the difference in time from the moment when the atom assumed the excited state $E_m$ to the moment when it made a transition to some other arbitrary state. The interval $\Delta t$ depends on the transition probability. If only spontaneous transitions are considered, the corresponding line width is called the natural width. Since the probability of spontaneous transitions between the Zeeman-split components is very small, the natural width of EPR lines is negligible; in the microwave region it is of the order of $10^{-4}$ c/s.

The natural line width does not take into account the interaction between the atom and its neighbors. Actually, in macroscopic systems there is spin-lattice interaction, i.e., the interaction between the paramagnetic ion and the thermal vibrations of the lattice. The thermal vibrations of the atoms in the lattice change the crystal fields to a certain degree, and as a result the value of the internal crystal field is not the same at different points and different moments of time. On the other hand, the interaction of the magnetic moment with the lattice (more accurately, with the crystal field of the lattice) affects the lifetime of the atom in a given state. Let us assume that the orbital angular momentum is suppressed (i.e., that the orbital levels are quenched) and that the spin vector is oriented in an external magnetic field. What factors could cause the establishment of the initial orientation? Any arbitrary interaction which could do this must be coupled magnetically with the spin magnetic moment of the electron. There are two types of such interactions: spin-orbit coupling and spin-spin coupling of adjacent electrons. Thermal fluctuations of the crystal field affect the orbital angular momentum and can cause a change in its orientation, while at the same time the orientation of the spin of the electron will also change. In exactly the same way the spin-spin interaction is dependent on the direction of interacting elementary magnetic dipoles and on the distance between them, so that thermal vibrations undoubtedly affect the spin-spin interaction. In both cases thermal lattice vibrations can cause transitions between the Zeeman sublevels.

A detailed computation of the "reorientation time" of the magnetic moment (spin-lattice relaxation time) is very complicated, and the expressions obtained from such computations agree qualitatively only with experiment. Moreover, it appears that energy exchange between the spin system and the lattice can take place in two ways: either directly via a whole quantum, equal to the quantum of energy of the lattice vibrations at the corresponding frequency (resonant process), or by scattering of a quantum of the lattice vibrations due to changes in its energy

(Raman scattering). Kronig has obtained the following expressions for the case $S = \frac{1}{2}$ [12, 40]:

$$\tau \sim \frac{\Delta^4}{\lambda H^4 T} \quad \text{(resonant exchange)} \tag{30}$$

$$\tau \sim 10^2 \cdot \frac{\Delta^6}{\lambda^2 H^2 T^7} \quad \text{(Raman scattering)} \tag{31}$$

where $\lambda$ is the spin-orbit coupling interaction constant, in $cm^{-1}$ (in some cases it is of the order of 800 $cm^{-1}$); $\Delta$ is the separation between the two lowest orbital levels, in $cm^{-1}$; $H$ is the strength of the external magnetic field, in ka/m; and $T$ is the absolute temperature.

The character of the dependence of $\tau$ on $\lambda$ and $T$ is understood. The larger the spin-orbit coupling, the easier it is for the crystal field to "reach" the spin and the shorter will be the lifetime of the atom. When the temperature increases, the thermal vibrations become larger and the changes in the crystal field more insignificant, causing a more probable "reorientation" of the spin and a decrease in the spin-lattice relaxation time. The character of the dependence of $\tau$ on $\Delta$ and $H$ is not so obvious and can be discovered only after detailed examination. It can also be remarked that at room temperature the predominant process is the scattering of the "lattice waves." This is explained by the fact that the lattice vibration frequencies and the Zeeman frequencies are completely different in magnitude; i.e., there are no conditions for resonant exchange. Conditions for resonant exchange appear only at very low temperatures (of the order of 4-20°K). Broadening of the EPR lines caused by spin-lattice interaction can be decreased by lowering the temperature, which sharpens the lines and amplifies the resonant peaks. In practice, the temperature is lowered until the line width is determined by spin-spin interaction and not by spin-lattice interaction. The broadening of the EPR lines due to spin-lattice interaction exceeds the natural line width by hundreds of times in magnitude and reaches hundreds of kilocycles and even tens of megacycles.

Spin-spin interaction also plays a large role. The nature of this interaction is such that each electron of the paramagnetic substance is under the influence of the internal magnetic field that arises as a result of alignment of the spin magnetic moments of all unpaired electrons of the substance. Due to this the value of the external magnetic field intensity increases by an amount $H_{ex}$ which is proportional to

$$H_{ex} \sim \mu \frac{1 - \cos^2\theta}{r^3} \qquad (32)$$

where $r$ is the distance between the adjacent dipoles; and $\theta$ is the angle between the direction of the external magnetic field and the directions of these dipoles.

Since this additional contribution has different values for different $r$ and $\theta$, various ions will be located in somewhat different fields. The resonance condition in this case becomes somewhat indeterminate and the spectral line is "smeared" by the amount

$$\Delta\nu = \frac{g\mu_0 H_{ex}}{h} \qquad (33)$$

Broadening of the lines due to the rotating component of the magnetic field vector also appears since the interaction of this component with other spins precessing with the same Larmor frequency induces transitions, i.e., decreases the lifetime of the ion in a given energy state.

The broadening of the line due to spin-spin interaction will be larger the larger the value of $H_{ex}$. The field $H_{ex}$ increases with the magnitude of the spin magnetic moment. Van Vleck has obtained the following expression for the mean-square line width [12]:

$$(\Delta H^2) = \frac{3}{4} g^2 \mu_0^2 S(S + 1) \sum_k \left( \frac{1 - \cos^2 \theta}{r^3} \right)_k^2 \qquad (34)$$

It follows from this formula that the only method of decreasing the spin-spin broadening is to increase the distance between the paramagnetic ions by diluting the investigated substance with a suitable nonmagnetic solvent. In this case the natural line width will not be determined by the interaction of the magnetic moments of the unpaired electrons of the neighboring paramagnetic ions but by the interaction with the magnetic moments of the nuclei of the neighboring diamagnetic atoms. If the paramagnetic ion is hydrated, i.e., surrounded by molecules of water, the indicated interaction with the magnetic moments of the nuclei (in this case protons) can be further decreased by growing the crystals in heavy water, since the magnetic moment of the dueteron is smaller than the magnetic moment of the proton. In any case, the spin-spin broadening can always be made negligible, i.e., such that it will not prevent the detection and study of EPR. It should be also noted that spin-spin interaction decreases rapidly with distance; therefore in computation of the line width only the influence of nearby ions need be considered. Furthermore, due to the angular function contained in Eq. (34), the spin-spin broadening depends on the orientation of the crystal in the external field.

The width of the line is also affected by exchange interaction. If the paramagnetic substance consists of ions of the same type, a fast exchange of electrons will tend to average out the effect of the internal fields, and this in turn will cause a decrease in the spread of the spectral line. For example, the line width of copper phthalocyanine due to strong spin-spin interaction should be about 30 ka/m, but exchange interaction decreases this width to 2.8 ka/m. If the exchange of electrons takes place between different atoms with different Larmor spin precessions, the exchange will tend to average out both precession frequencies and will thus cause a broadening of the line.

The exchange interaction strongly depends on the distance between the atoms, and its effect can be decreased by increasing the distance between the paramagnetic ions, i.e., in the same way as is done to decrease the effect of spin-spin interaction.

The broadening of the line can also be caused by saturation. The saturation effect causes a decrease in intensity at the absorption maximum and an increase in the line width (Section 4). It should be noted, however, that in EPR experiments the broadening of the line due to saturation almost never takes place because the power levels used in practice never reach a value that would cause the line broadening due to saturation to be larger than the broadening due to other effects.

Finally, let us note that line broadening can also occur due to an inhomogeneous magnetic field. It is completely independent of the properties of the investigated paramagnetic substance and is caused by differences in the external magnetic field in various parts of the paramagnetic substance, so that resonance in various parts of the paramagnetic substance will occur at different frequencies. The inhomogeneity of the magnetic field is caused by several factors: bad machining of the surface of the magnet pole faces, bad parallelism of these surfaces or their small size, so that edge effects become significant. In the case of EPR for observation of narrow lines (of the order of 10 amp/m), magnetic fields which are homogeneous to $10^{-5}$ are sufficient.

## 9. APPLICATIONS OF ELECTRON PARAMAGNETIC RESONANCE

*Determination of nuclear constants.* The interaction of the magnetic nuclear moment with the internal atomic magnetic field $H_A$ leads to the hyperfine structure of the atomic energy levels. The direction of the field $H_A$ becomes the quantization axis for the nuclear magnetic moment, with the result that each level of the fine structure splits up into $2I + 1$ sublevels, creating a hyperfine multiplet ($I$ is the spin of the nucleus). The separation between the components of the hyperfine multiplet is equal within the first-order approximation.

Since the external magnetic radio frequency field affects only the moment of an unpaired electron, in electronic transitions the orientation of the nuclear magnetic moment is conserved; i.e., the selection rule $\Delta M_I = 0$ is satisfied. The transitions which correspond to this rule are shown in Fig. 24. It is clear that for the case under consideration $\left(\text{spin } I = \frac{3}{2}\right)$ the EPR line splits into four components.

The hyperfine structure consists of $2I + 1$ equally spaced components, so that paramagnetic resonance is an ideal method for determining nuclear spins. In cases when it is possible to compute the internal atomic magnetic field $H_A$ at the location of the nucleus, it is also possible to find the value of the nuclear magnetic moment and the nuclear gyromagnetic ratio. Since the computation of the field $H_A$ is very difficult, however, the method of determining the magnetic nuclear moments from the hyperfine structure of the EPR spectrum is significantly less accurate than the nuclear resonance method. In the case when the nuclear magnetic moment is already known, from measurement of the hyperfine structure of the EPR spectrum one can obtain an accurate value of the internal atomic field, which is an important characteristic of the electron configuration of an atom or an ion.

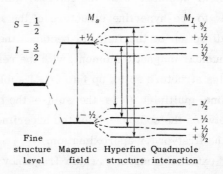

FIG. 24. Hyperfine structure of Zeeman sublevels

The above considerations, from which the fact of equal spacing between the lines of the hyperfine structure can be deduced, are valid when the external magnetic field is much stronger than the field of the nucleus itself. If these fields are of comparable magnitude, the hyperfine structure sublevels will contain admixtures of different quantum states $M_I$ which will cause the separations between the sublevels to vary. Distinguishing properties of this effect are the smooth changes of separation between the sublevels and the disappearance of the effect in strong magnetic fields. Investigation of the monotonic change of distance between the maxima in a number of cases yields supplementary information on the magnitude and nature of the forces characterizing the interaction between the nucleus and the shell as well as the external magnetic field.

There is one more reason for unequal separations between the sublevels of a hyperfine multiplet. If the nucleus possesses an electric quadrupole moment, interaction of the crystal field with this moment will take place, and since the quadrupole moment is connected through the magnetic moment of the nucleus, the crystal field shows an effect on the energy sublevels of the hyperfine structure (Fig. 24). In general, the result of the quadrupole interaction depends on the relative orientation of the external magnetic field and the internal crystal field. If the magnetic field is parallel to the axis of symmetry of the internal crystal field, the displacement of all energy levels is the same, and consequently the frequencies of the observed transitions do not change. In the case when there is an angle between the axis of the crystal field and the direction of the magnetic field, the displacement of levels may not be identical and the distance between the components of the hyperfine structure at the ends of the spectrum becomes greater than at the center. Due to quadrupole interaction the transitions with $\Delta M_I = \pm 1, \pm 2$ become allowable. The result is that every peak of the hyperfine structure will have quadrupole satellites of smaller intensity. The intensity and number of these satellites

depend very strongly on the angle between the fields. By changing this angle one can study the quadrupole interaction, in rela-
tion to the separation between the components of the hyperfine
structure, with respect to the number and intensity of the
quadrupole satellites. The theory of EPR leads to the conclusion that the intensity of the "forbidden" lines relative to the
intensity of the main lines of the hyperfine structure is of the
order of $Q'^2/K$, where $K$ is the splitting between the hyperfine
structure levels, and

$$Q' = \frac{3e}{4I(2I - 1)} \frac{\partial^2 V}{\partial z^2} Q$$

$Q$ being the electric quadrupole moment of the nucleus.

Thus it can be said that the intensity of quadrupole satellites
is proportional to $e \dfrac{\partial^2 V}{\partial z^2} Q$ (c/s). This quantity is called the
quadrupole interaction constant, and it is this constant that is
determined in experiments. If the distribution of the crystal
field at the location of the quadrupole moment were known with
sufficient accuracy, $Q$ could be determined with great accuracy.
But since the quantity $\dfrac{\partial^2 V}{\partial z^2}$ cannot be calculated easily, the
accuracy in determination of the quadrupole moment of the
nucleus does not exceed two significant figures. On the other
hand, if the magnitude of the quadrupole moment of the nucleus
is known, from the displacement of the second-order levels and
from "forbidden" transitions one can determine the gradient
of the electric field which interacts with the quadrupole moment.

*Study of transition elements.* The transition elements were
the very first field to which the EPR method was applied. A

characteristic property of the atoms of the transition elements is their unfilled inner electron shells. For example, the elements of the iron group have an incompletely filled $3d$ shell, the elements of the palladium group an incomplete $4d$ shell, and the rare earth elements an incomplete $4f$ shell.

There are two basic directions in which studies of the transition elements are pursued by the EPR method. The first includes the analysis of the position, the splitting and the intensity of the spectral lines. Such information reveals much about the nature of interatomic interactions and thus the nature of the chemical bond, which is of great value in the development of the theory of magnetism and the theory of the solid state. The paramagnetic resonance method has a great advantage in this respect because it allows one to extract one or two levels for detailed study. In this way it is possible to obtain accurate data on each level separately. Such data form excellent material for verification of various theories. It should also be noted that in some cases the old theories did not hold out under this check and required refinements and changes (and in some cases they were shown to be inapplicable). A good example of this is the $Mn^{++}$ ion. Its ground state is $^6S$ ($S = 5/2$, $L = 0$), orbital degeneracy is absent, and there is a sixfold spin degeneracy.

The earlier theory stated that the only mechanism of splitting of the ground state level was that of splitting in a cubic field. The result was that in a cubic field the sixfold spin degeneracy was removed and two levels were created: a doublet and a quadruplet. After measurement of the paramagnetic absorption, however, it became clear that the old theory is completely insufficient for explanation of the EPR spectrum of this ion. A completely new mechanism of the splitting of the basic term of the $S$ state was proposed to explain the experimental facts. The $S$ state is characterized by the fact that it has a spherically symmetrical electron cloud. The energy levels for various values of $M = \pm 5/2, \pm 3/2, \pm 1/2$ are identical (sixfold spatial spin degeneracy). But if it is assumed that for any reason this symmetry is violated, we immediately find that the energy of the levels with $M = \pm 5/2$, $M = \pm 3/2$ and $M = \pm 1/2$ will be different. Thus the perturbation of the electron cloud causes the ground state term to split into three doublets. In an external magnetic field each of these doublets splits up into two levels, or a total of six levels, between which transitions can occur in accordance with the selection rule $\Delta M = 1$. This model completely agreed with experiment. The cause of perturbation of the electron cloud is the dipole-dipole interaction between the magnetic moments of the electrons.

The second direction in studies of the properties of the transition elements by the EPR method centers on the analysis of the hyperfine structure of the electron states, which allows formulation of theories on the properties of the nucleus. The spins and the magnetic and quadrupole moments of the nuclei of the transition elements were measured with the help of EPR. In one case it was the first measurement of these quantities (for example, spins and magnetic moments of nuclei of $Nd^{143}$ and $Nd^{145}$, spins of nuclei of $Sm^{147}$ and $Sm^{149}$), and in the other case it was a refinement of previously obtained values. An extremely valuable set of data on the so-called configuration interaction was obtained for the iron group. The true electronic structure of the paramagnetic ions of the iron group was discovered from these data. It became evident that, due to configuration interaction, one electron from the $3s$ state was excited to the $4s$ state, a fact which is true for most of the elements of the iron group.

*Other applications of EPR.* In recent time EPR methods have been widely used in studies of free radicals. Free radicals are compounds which have one or several unsaturated bonds. As a result of the unsaturated chemical bond there appear electrons with unpaired spins, whose presence assures the appearance of EPR. In a majority of cases the free radicals are very unstable and their lifetime is very brief. At the same time, free radicals are present in almost all living tissues and determine significantly the chemical and biophysical processes taking place in those tissues. Therefore a knowledge of the nature and of the pecularities of behavior of free radicals is of great interest in biophysics, biology and medicine. Investigation of free radicals by the EPR method is conducted on a very wide scale at the present, and major and interesting discoveries can be expected in this field.

EPR methods are also useful and effective in studies of properties of matter subjected to radiation. During radiation some defects appear in matter so that electrons accumulate at these places (creating $F$ centers, for example). These trapping centers possess paramagnetic absorption and therefore can be studied by EPR methods.

EPR methods can also be used for qualitative and quantitative analysis of any paramagnetic substance.

The phenomenon of electron paramagnetic resonance can also be used for measurement and stabilization of magnetic fields. Let us assume that DPPH is chosen as the standard substance. It follows from the conditions of resonance that $H = 2.837 \times 10^{-5} \nu$, where $H$ is measured in amp/m and $\nu$ is in c/s. A change of frequency in the limits from 1 to 1000 Mc/s allows one to determine a field from 0.05 ka/m to 10 ka/m. For larger values of $H$ the use of nuclear magnetic resonance is advisable. The lower limit is determined by the line width of the utilized substance. The width of the line of DPPH is of the order of 0.15 ka/m, and furthermore the $g$ factor of DPPH is anisotropic for small fields (below 0.25 ka/m), so that DPPH should not be used for measurement of small fields. Other substances could be used (there are substances with a line width of the order of 1.5 amp/m), but in the region 0-0.15 ka/m there is another, more powerful method, based on utilization of the decaying precession of the nuclear magnetization vector around the direction of the measured field.

## Chapter IV

## Nuclear Magnetic Resonance

### 10. THE PHENOMENON OF NUCLEAR MAGNETIC RESONANCE

If the spin of a nucleus $I$ is different from zero and if the nucleus possesses a magnetic moment then, on application of an external magnetic field, the orientations of the magnetic moment of the nucleus will be quantized and corresponding energy levels will be formed. An alternating magnetic field at resonant frequency will cause transitions between these levels. In a way analogous to electron paramagnetic resonance, the condition for resonance can be written in the form

$$h\nu = g_n \mu_{n0} H \tag{36}$$

where $\mu_{n0}$ is the nuclear magneton; and $g_n = \mu_n/\mu_{n0}I$ is the nuclear $g$ factor, or the splitting factor.

The nuclear magneton is 1/1837 times smaller than the electron (Bohr) magneton, and therefore the frequencies of transitions caused by nuclear magnetic moments fall in the region of several megacycles in fields of hundreds of kiloamperes per meter, while the transitions caused by electron magnetic moments in the same magnetic fields fall in the microwave frequency region. Thus, for example, for a magnetic field of 400 ka/m the NMR resonant frequency for protons is of the order of 20 Mc/s [13].

The phenomenon of nuclear magnetic resonance has much in common with electron paramagnetic resonance. The selection rules for NMR transitions are analogous to the selection rules for the EPR case: $\Delta M = \pm 1$. Just as in the case of EPR, a weak magnetic field oscillating at the resonant frequency and perpendicular to the constant magnetic field is required for observation of NMR. Just as in observations of EPR, NMR can only be observed in macroscopic systems, when energy transfer between the nuclear spin system and matter is possible. If the energy received by the spin system could not be transferred to matter, saturation would occur and a steady-state absorption of the energy of the alternating magnetic field could not be observed.

In Chapter II general laws governing the phenomenon of magnetic resonance were examined. The expressions obtained there are fully applicable when the paramagnetism of the substance is caused by nuclear magnetic moments. The mechanism of longitudinal and transverse relaxation in the case of NMR, however, has several characteristic properties.

The most realistic mechanism of spin-lattice relaxation was proposed by Bloembergen, Purcell and Pound [24]. Their initial assumption was that every nucleus is located in an internal local magnetic field $H_{loc}$. Due to Brownian motion of particles, elastic vibrations, diffusion, etc., the field $H_{loc}$ will change continuously with time. Thus the local magnetic field is a fast-changing function of time and can be represented by a Fourier series. The most interesting terms of this series are those whose frequency is the same as the Larmor frequency of precession of the nuclear magnetic moments. When these frequencies are identical, resonant interaction between the perpendicular portion of the Fourier component of $H_{loc}$ and the nuclear magnetic moment takes place and the moment is reoriented. The energy from the nucleus is transferred to the fluctuating magnetic field $H_{loc}$ and this, in turn, causes an increase in the thermal motion of the atoms.

In liquid substances without paramagnetic atoms or ions, the main factor which causes the appearance of internal magnetic fields $H_{loc}$ is a random (translational and rotational) thermal motion of the nuclei. The energy of spins is converted into the energy of the Brownian motion. At the same time, apparently, internal molecular interactions predominate over the interactions between the molecules. Due to this the probability of energy transfer by the spin to its own molecule is greater than that of transfer to an adjacent molecule.

In metals the main factors establishing the field $H_{loc}$ are, of course, the conduction electrons which pass close to the nucleus. Here the energy of the nucleus is used up to increase the kinetic energy of the electrons.

In the case of nonmetallic solids, all of whose electronic configurations are diamagnetic, the factors which lead to the establishment of $H_{loc}$ can be the rotational Brownian motion of atoms or molecules, diffusion and lattice vibrations. The studies of NMR in solids have shown, however, that the lattice vibration is immaterial and that Brownian motion and diffusion in solids are negligible except in a few solids (solid hydrogen, solid cyclohexane, metals). In a majority of cases the above causes of establishment of $H_{loc}$ cannot yield relaxation times $\tau_1$ which are observable in practice. Therefore a new and more powerful mechanism of spin-lattice interaction was proposed by Rollin and Hatton [25]. Their main idea is that solid substances completely deprived of paramagnetic impurities (in the form of paramagnetic ions or $F$ centers) simply do not exist. Theoretical and experimental work has shown that as little as 0.0001% of impurities is a decisive factor in the process of spin-lattice interaction. It is important to note that the mechanism of the energy transfer from the nuclear spins to the lattice can have two forms. If the concentration of paramagnetic ions is higher than 1%, the spin-exchange processes between the nuclear spins and the spins of the ions predominate. In this case the energy of the nuclear spin is spent to change the frequencies of precession of the ion spin. A transfer of energy takes place from the nuclear spin system to the spin system of the ions, and then from the spin system of the paramagnetic ions to the lattice. It must be remembered that the spin-lattice relaxation time for ions is very small ($10^{-4}$-$10^{-6}$ sec) compared to the time of nuclear spin-lattice relaxation ($\sim 1$ sec). If the concentration of paramagnetic ions is below 1%, the spin-diffusion process also becomes significant when the Brownian and the diffusive motions of the ion spins take place and give rise to $H_{loc}$. In this case the energy of the nuclear spin system is converted directly into thermal energy of the lattice.

From the mechanism of spin-lattice relaxation proposed by Bloembergen, Purcell and Pound, there follow two interesting conclusions. Since $H_{loc}$ is determined by the Brownian motion of atoms and molecules, there should exist a connection between the spin-lattice relaxation time $\tau_1$ and the viscosity. The higher the viscosity, the smaller the Brownian motion, so that the spectrum of random changes in locations of particles (and therefore the spectrum of $H_{loc}$) is displaced toward the low-frequency region. The resonant interaction of $H_{loc}$ with the spin system is thereby increased and the relaxation time decreased. A further increase in viscosity will cause a decrease in Browian motion and a decrease in spin-lattice interaction, i.e., an increase in $\tau_1$. Thus a plot of $\tau_1$ as a function of viscosity should have a minimum. Experimental investigations have supported this speculation.

The second conclusion is that the addition of paramagnetic ions to matter to create significant magnetic fields should drastically increase the spin-lattice interaction and decrease the time $\tau_1$. This fact is of great importance for experimental observation of NMR. A large value of $\tau$ causes fast saturation and makes observation of the NMR absorption impossible. Therefore, in a number of cases a paramagnetic impurity is used purposely to decrease $\tau_1$.

The mechanism of transverse relaxation, based on interactions in the nuclear spin system, is in general analogous to the process of transverse relaxation in the case of electron paramagnetic resonance. In a nuclear spin system, just as in an electron system, there exist spin-exchange processes and magnetic dipole-dipole interactions between the magnetic moments of the nuclei. The dipole-dipole interaction causes the precession frequencies of various nuclei to differ somewhat from one another; as a consequence of this, the resonant frequency is "smeared"; i.e., the spectral line is broadened. A similar effect could be caused by the inhomogeneity of the external magnetic field and by the diamagnetic shielding of the nuclei by the electrons in the atomic shells. The phase coherence will also be destroyed by the interaction of the nucleus with neighboring atoms or molecules if these neighbors are different from one another. This effect will be most pronounced in viscous substances, in which the motion of the molecules is insignificant. In low-viscosity liquids the random motion of the molecules can become faster than the precession of the magnetic moments of the nuclei. In this case the fluctuations of the local magnetic fields will average out and the transverse relaxation time $\tau_2$ will increase.

Since the interaction of the nuclear spin system with the lattice is very weak and the spin-spin and spin-exchange processes are also much weaker than in the case of EPR, the NMR lines in the absence of saturation should be very narrow and comparatively weak. Countless experimental data show that characteristic properties of nuclear magnetic resonance lines are actually their narrowness and low intensity. This places some specific requirements on the methods of observing NMR. In particular the narrowness of NMR spectral lines requires more stringent limits on the inhomogeneity of the constant magnetic field than in the case of EPR. For field intensities of the order of several hundreds of ka/m the instability of the magnetic field should not be more than several tenths of amp/m. Very frequently, due to inhomogeneity of the magnetic field in experiments, it is impossible to measure the width of the line resulting from spin-lattice or spin-spin interaction because this line is completely determined by the inhomogeneity of the magnetic field. In some cases the sample under investigation is rotated to decrease the line-broadening effect due to inhomogeneity of the field. This averages the field inhomogeneity in the direction perpendicular to the axis of rotation.

## 11. METHODS OF OBSERVING NMR

The methods for observation of nuclear magnetic resonance can be arbitrarily divided into the following four groups: 1) methods which use a marginal oscillator [9, 13, 26], 2) methods which use bridge circuits [9, 13, 27, 32], 3) pulse methods [9, 13, 33] and 4) methods based on nuclear induction [9, 19, 13].

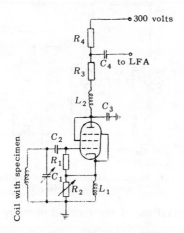

FIG. 25. Autodyne circuit of Hopkins [34]

A classical example of the first group is the autodyne scheme of Hopkins (Fig. 25). The investigated sample is placed in the grid circuit coil of an oscillator. A constant magnetic field $H$ is applied perpendicularly to the alternating magnetic field. When the absorption line is traversed, the $Q$ of the circuit changes, causing a change in the amplitude of the oscillations. These changes can be amplified and recorded.

The oscillator is actually an amplifier with positive feedback and can be regarded as a negative input impedance device, with the negative impedance in parallel with the tuned circuit. The magnitude of the negative impedance is adjusted to make the amplitude of the oscillations very small. This makes the circuit very sensitive to small changes in the $Q$ of the tuned grid circuit caused by resonant absorption. Besides having a high sensitivity, this method is also good from the point of view of saturation, which does not become a great problem at small amplitudes.

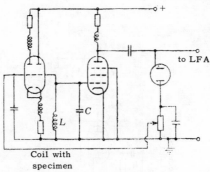

FIG. 26. Marginal oscillator circuit of Pound for observation of NMR [9]

A second variation of the method of marginal oscillators is the method of Pound (Fig. 26). The inductor $L$, in which the investigated sample is placed, and the capacitor $C$ make up the tuned circuit of the oscillator. The oscillations are amplified by the second tube and after detection are fed to a low-frequency amplifier. In Pound's scheme, high-frequency amplification is used before detection; Hopkins' method uses detection at the grid of the oscillator tube. Grid rectification with the same tube has the advantage that the amplitude of oscillations is maintained at a low level (since an increase in the amplitude of oscillations increases the negative resistance at the grid, and this in turn decreases the gain). On the other hand, amplification at high frequency (Pound's scheme) allows one to lower the noise level and improve the signal-to-noise ratio.

Since the interaction of the nuclear spin system with the lattice in a number of cases is very weak, it is desirable that the amplitude of oscillations of the RF magnetic field be as small as possible. The methods described above do not allow one to obtain a voltage smaller than 0.1 volts across the coil and maintain stable operation of the oscillator at the same time. This disadvantage is partially removed in a radio frequency spectrometer proposed by Lemanov [26]. The oscillator circuit is shown in Fig. 27. By a suitable choice of tube parameters

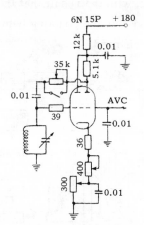

FIG. 27. Lemanov's marginal oscillator circuit for observation of NMR [26]

and operating conditions, minimum voltages of 0.02 volts were achieved in the oscillator circuit. The frequency of oscillation ranged from 2 to 13 Mc/s.

The second popular group of methods for observation of NMR are the methods utilizing bridge circuits. As an example, Fig. 28 shows the apparatus used by Bloembergen, Purcell and Pound. Two almost identically tuned circuits, the coil of one of which contains the investigated specimen, are fed in parallel by an RF oscillator. From the tuned circuits the signals are fed to point $A$ and one is subtracted from the other, since between points $A$ and $B$ there is a cable whose length is equal to one half-wavelength. The bridge is balanced in phase and amplitude by means of condensers $C_1$ and $C_4$. Tuning of $C_1$ affects the phase balance (because the characteristic impedance of the cable $R_2$ is small), so that condenser $C_1$ is in effect ganged together with the condenser $C_6$, whose capacitance decreases when $C_1$ is increased, with the result that the tank circuit tuning remains constant and only the coupling between the tank circuit and the oscillator is changed.

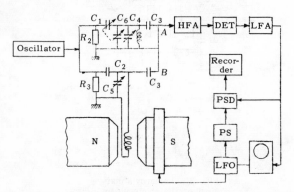

FIG. 28. Apparatus of Bloembergen, Purcell and Pound for observation of NMR [16]

The RF bridge is usually not completely balanced. The character of any residual unbalance must be known, however, since it determines the shape of the NMR signal. An absorption

signal will be observed with amplitude unbalance, and a dispersion signal will be observed with phase unbalance. The situation
here is completely analogous to the case of observation of EPR
using radio frequency spectrometers with waveguide bridges
(Section 6).

There are many variations of RF bridge circuits that can
be used in radio frequency spectrometers for observation of
NMR. Their main advantages are: first, it is possible to obtain
significantly smaller voltages across the working coil than in
the case of the marginal oscillator; second, a decrease of the
initial voltage at the amplifier input causes an increase in the
depth of amplitude modulation of this voltage by the NMR signal
and enables a greater high-frequency amplification; third, a
large portion of the noise that modulates the input voltage of the
oscillator is cancelled.

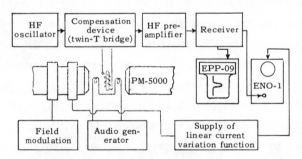

FIG. 29. Radio frequency spectrometer for observation of the fine
structure of NMR spectra [27]

To complete the discussion of the methods for observing
NMR in which bridge circuits are used, we will familiarize
ourselves with one more spectrometer [27], which is especially
designed for observation of the fine structure of NMR spectra
(Fig. 29). The spectrometer uses a permanent magnet; the

field in the magnet is 360.5 ka/m, the diameter of the pole faces is 220 mm, and the air gap is 32 mm. The deviations from flatness of the pole faces are less than 0.5 microns and the relative inhomogeneity of the field in the central region is not greater than $2 \times 10^{-6}$ in a volume of 1 cm$^3$.

The investigated specimen, contained in a thin glass ampule, is placed in the region of the most homogeneous magnetic field by means of a special coordinate positioning mechanism. To obtain an averaging effect of the longitudinal component of magnetic field, the glass probe is rotated by an air turbine at a velocity of 10,000 r/m.

An outstanding property of this radio frequency spectrometer is that it uses a heterodyne detection system. The receiver is a superheterodyne amplifier with one IF amplification stage at 110 kc/s. The receiver gain is $\sim$ 10,000 and the bandwidth is 4 kc/s. The frequencies of the local oscillator and the HF oscillator are quartz-stabilized so that the relative frequency stability during the time required to record the NMR spectrum is not worse than $10^{-7}$. The radio frequency spectrometer allows either visual examination of the NMR spectrum on an oscilloscope screen (which may be photographed), or recording of the NMR spectrum with graphic recorder EPP-09.

Pulse methods for observing NMR are also of great importance. Their characteristic property is that modulation of the magnetic field during the time of observation of the NMR spectrum is not required, because an RF magnetic field of resonant frequency is supplied as a train of pulses of predetermined width. The transient signals of nuclear induction are observed during the time when the pulse is acting and also after it stops.

Let us investigate what happens to the magnetization vector $\vec{M}$, which rotates around the direction of the constant magnetic field $\vec{H}$, when it is perturbed by an RF pulse. We will introduce a rotating coordinate system whose $z$ axis coincides with the direction of the uniform magnetic field and whose frequency of rotation is the same as the frequency of precession of the vector $\vec{M}$. In the rotating coordinate system the vector $\vec{M}$ will be stationary. Let us now superimpose an alternating magnetic field at the resonant frequency, i.e., rotating at the same speed as the rotating coordinate system. Suppose that the amplitude of this field is equal to $H_1$ and that its direction is along the $x$ axis of the rotating system (Fig. 30). The interaction between the field $H_1$ and the magnetic moment causes precession of vector $\vec{M}$ around $H_1$ with a precession frequency $\omega_1 = \gamma H_1$. If the pulse length $t_1$ is such that the condition $\omega_1 t_1 = \pi/2$ is satisfied, the vector $\vec{M}$, as a result of precession, will rotate by 90° and point along the $y$ axis of the rotating coordinate system. After the pulse has ended, the magnetization vector $\vec{M}$ is only subject to perturbation by the uni-

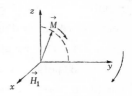

FIG. 30. Rotation of the resultant nuclear magnetization vector from its steady-state position under the influence of an RF pulse

form magnetic field $H$ (if relaxation processes are neglected) and completes the precession around this field (in the stationary coordinate system). As a result an emf is induced in the coil. This signal, however (signal of free precession), is monotonically decreasing. Due to inhomogeneities of the external and internal magnetic fields the precession frequencies of various nuclei are different. As a result the total magnetization vector $\vec{M}$ starts to decompose into separate vectors and a spreading "fan" of magnetization vectors is created (Fig. 31). Consequently the emf induced in the coil is damped out.

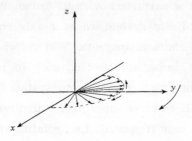

FIG. 31. Spreading of the resultant nuclear magnetization vector after the first RF pulse has ended [33]

The spreading fan of the magnetization vectors can be again compressed together. A rotation by 180° around the axis is required for this. Since the direction of the spreading of the vectors remains the same, it now results in a compression (Fig. 32). The required 180° rotation can easily be realized by a second pulse of amplitude $H_2$ whose length $t_2$ satisfies the condition $\omega_2 t_2 = \pi$ (where $\omega_2 = \gamma H_2$). The compression of the fan of magnetization vectors into one vector will cause the appearance of an emf or, as it is called of a spin-echo signal.

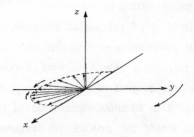

FIG. 32. Compression of the resultant nuclear magnetization vectors under the influence of the second RF pulse [33]

If before the rotation of the fan by the second pulse the fan was spread over a certain angle during the time $(\tau - t_1)$, then after the rotation it will be compressed during the same time $(\tau - t_1)$ after the end of the second pulse. Therefore the time from the beginning of the first pulse to the maximum of the echo signal is $2(\tau - t_1) + t_1 + t_2 = 2\tau - t_1 + t_2$ (Fig. 33). If the length

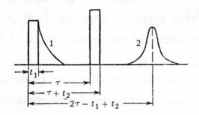

FIG. 33. Appearance of the free precession signal (1) and the spin echo (2) as a result of the effect of the RF pulses [33]

of both pulses is the same $(t_1 = t_2)$, the spin-echo signal appears at the time $2\tau$ ($\tau$ is the interval between pulses). Figure 34 shows a block diagram of a system for observation of spin-echo

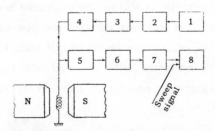

FIG. 34. Block diagram of a system for spin-echo observation [36]. 1—Pulse generator; 2—control circuit (gate); 3—modulator; 4—power amplifier; 5—high-frequency amplifier; 6—limiter and detector; 7—low-frequency amplifier; 8—oscilloscope

signals [36]. It should be noted that the method of observation of spin-echo signals has much in common with the methods used in radar. As is evident from the figure, the system consists of two channels: transmitting and receiving. The transmitting channel consists of a pulse generator 1, control circuit 2, modulator 3, and power amplifier 4. The receiving channel consists of the high-frequency amplifier 5, detector with limiter 6, low-frequency amplifier 7, and oscilloscope 8. The limiter 6 protects the system from large-amplitude pulses. The control circuit 2 generates pairs (or triplets) of pulses in a continuous succession. The oscilloscope sweep starts at the beginning of the first pulse.

The spin-echo signals are observed in the interval between the RF pulses, i.e., at moments of time when the RF generator is not working. Due to this the sensitivity of the system is primarily determined by the noise in the amplifier (with other methods the main source of noise is the high-frequency oscillator itself).

The above method for observing NMR, based on the appearance of spin echoes, is obviously not unique. However, other variants of pulse methods will not be examined here.

One of the first methods devised for observation of NMR was the method of nuclear induction. It was described by us previously in conjunction with the methods for the study of electron paramagnetic resonance (Section 6) and will not be examined here.

## 12. NUCLEAR QUADRUPOLE RESONANCE

If a nucleus has a spin larger than 1/2, it possesses an electric quadrupole moment (for example, nuclei of D, N, B, Al, S, Cl, Br). The quadrupole moment $Q$ interacts with the internal electric fields present in matter. This interaction is especially large in solids, in which strong internal crystal fields exist. As a result of quadrupole interaction, the nucleus is orientated with respect to the internal electric field. At the same time the magnetic moment of the nucleus is oriented and the spatial degeneracy of the energy levels is removed. The quantum-mechanical computation gives the following expression for the hyperfine structure levels:

$$E_{M_I} = \frac{1}{4} e^2 Qq \frac{3M_I - I(I + 1)}{I(2I - 1)} \tag{37}$$

where $q = \dfrac{\partial^2 V}{\partial z^2}$; $z$ is the axis of symmetry of the electric field.

It is directly evident from this expression that the separation between the sublevels with quadrupole splitting is not constant and increases with an increase in $M_I$ (Fig. 35). Moreover, each

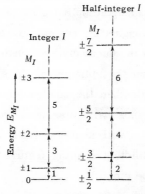

FIG. 35. Energy levels of a nuclear electric quadrupole in an axially symmetric electric field (unit of

energy is $\dfrac{3 e^2 Qq}{4I(2I - 1)}$)

quadrupole-split level has a twofold degeneracy which can be removed by means of an external magnetic field. In a number of cases the nuclear quadrupole splitting is quite large and transition frequencies are on the order of several tens of mega-cycles. Thus it is possible to obtain the spectrum of nuclear resonant absorption in the absence of the constant magnetic field.

Nuclear quadrupole resonance is frequently called pure quadrupole resonance [9, 13]. Both of these names, however, do not reflect one important property of the phenomenon, which is that this resonance is still magnetic, since the interaction between the nucleus and the high-frequency magnetic field which induces nuclear transitions is essentially magnetic. The difference between quadrupole resonance and the usual resonance is that the splitting of the levels is not caused by an external magnetic field but by the internal crystal field.

When a uniform external magnetic field is applied the double degeneracy of the quadrupole splitting sublevels is removed and the lines of "pure magnetic" resonance appear. At the same time the intensity of the quadrupole resonance lines decreases. In situations in which the splitting of lines in a magnetic field is significantly larger than the quadrupole splitting, the quadrupole resonance lines will disappear completely, and the quadrupole interaction will manifest itself as a shift in levels. Due to this the equidistant spacing feature of the magnetically split sublevels will be violated.

The line width of pure quadrupole resonance is usually $10^{-2}$-$10^{-5}$ times the resonant frequency. The main causes of line broadening are: magnetic dipole interaction between neighboring nuclei, broadening due to short relaxation time $\tau_1$, crystal lattice defects due to which the gradient of the crystal field varies somewhat from one molecule to another, and torsional vibrations of molecules which modulate the gradient of the electric field. It is interesting to note that torsional vibrations decrease the average value of the gradient of the crystal field and thus decrease the resonant frequency.

Methods for observing nuclear quadrupole resonance differ somewhat from the methods used for observing NMR. First, the applied uniform magnetic field is absent and the transition frequencies are determined by the internal electric-field gradient in the crystal. Therefore, in order to find a resonant line it is necessary to change the frequency of the high-frequency electromagnetic radiation. Furthermore, the effect of quadrupole resonant absorption in the majority of cases is very weak, leading to low values of signal-to-noise ratio. Thus the required apparatus must assure very large amplification and reliable extraction of very weak signals from noise.

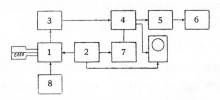

FIG. 36. Quadrupole radio frequency spectrometer [37]. 1—Oscillator-detector; 2—sinusoidal voltage generator; 3—narrow-band amplifier; 4—phase-sensitive detector; 5—amplifier; 6—recorder; 7—frequency doubler; 8—slow sweep of the oscillator frequency

Figure 36 is a block diagram of a quadrupole radio frequency spectrometer [37]. The investigated specimen is placed in the tank circuit coil of the oscillator-detector 1. The sinusoidal voltage generator 2 modulates the frequency of the oscillator-detector. After detection, the nuclear quadrupole resonance signal goes through the narrow-band amplifier 3, which is tuned to the second harmonic of the signal (twice the frequency of the modulation voltage). From the amplifier the signal is fed to a phase-sensitive detector 4 and then through amplifier 5 to recorder 6. The reference voltage for the synchronous detector is obtained from frequency doubler 7. The block diagram also shows a device for slowly sweeping the frequency of oscillator-detector 8, used to locate the resonance. It is also possible to observe the signal on an oscilloscope. The observation is made at the second harmonic because with sinusoidal modulation the resonance condition is satisfied twice during each modulation cycle. Due to this the spectrum of the signal is dominated by the component at twice the modulation frequency.

The sensitivity of quadrupole radio frequency spectrometers is comparatively low (usually $\sim 10^{-3}$ of a gram-equivalent for I, and $10^{-1}$ for Cl). During observation by the above method of nuclear quadrupole resonance in sodium chlorate $NaClO_3$ with a sample volume of 3 cm$^3$ and a filling factor of $\sim 0.6$, a signal-to-noise ratio of $\sim 100$ was achieved. The minimum concentration of $Cl^{35}$ nuclei required for good registration of a signal with a line width of 1 kc/s is $10^{21}$ per cubic centimeter.

The quadrupole moments of the nuclei and the electric field at the nuclear position differ in magnitude to a great extent, so that the nuclear quadrupole resonance lines of various elements differ widely in frequency from one another. Thus, observed transition frequencies for $N^{14}$ nuclei are in the region 1-3 Mc/s, for covalent bonds of $Cl^{35}$ in the region 30-70 Mc/s, for $Br^{79}$ in the region 270-300 Mc/s, and for $I^{127}$ in the region 1500-3000 Mc/s. It is obvious that this frequency spread cannot be covered with one oscillator. Therefore, a quadrupole radio-frequency spectrometer for the nitrogen region, which is of great interest in biological investigations, will be substantially different from a spectrometer for the bromine or iodine region, required for studies of metal-organic compounds.

## 13. APPLICATIONS OF NMR

*Determination of nuclear constants.* One of the main applications of nuclear magnetic resonance is in measurement of nuclear magnetic moments. The measurement is based on Eq. (36). The accuracy of measurement of $\mu_n$ will be determined by the accuracy with which the four quantities $h, \nu, I$ and $H$ are known. It is found that the limit of accuracy is determined by Planck's constant, which is known with an accuracy of $10^{-4}$. In relative measurements, when the nuclear magnetic moments are only compared, the accuracy can be substantially increased. Since the frequency can be measured with an accuracy exceeding $10^{-6}$, the ratio of magnetic moments can be measured with a higher accuracy. In situations when reliable data on the magnitude of the spin are lacking, instead of determining the magnetic moment one is limited to the determination of $g_n$, the nuclear $g$ factor (the quantity $g_n I$ is the value of the nuclear magnetic moment in nuclear magnetons).

The determination of nuclear spins with the help of NMR is accomplished indirectly, using second-order effects. There are two basic ways to determine the spin. The first is based on utilization of quadrupole interaction. As was indicated previously, quadrupole interaction causes displacement of energy levels. As a result of this, instead of one absorption line one will observe $2I$ lines, so that the number of components will determine the spin of the nucleus.

If quadrupole interaction is absent the spin may be determined from measurements of the absolute intensity of the line.

The fact is that the line intensity depends on how many separate transitions contribute to the total intensity. In order to determine spins by this method, relative measurements of intensities are usually conducted. Some of the first measurements of this type were used to determine the spin of the tritium $H^3$ nucleus. In the experiment, signal intensities from a sample containing hydrogen isotopes $H^1$ and $H^3$ in a known proportion were compared; it was shown that both isotopes have the same spin. It is significant that in such measurements great accuracy is not required because it is known in advance that the spin is either an integer or half-integer.

From measurements of nuclear quadrupole resonance one can obtain data on nuclear quadrupole moments and electric field gradients at the positions of the nuclei. It may be seen directly from expression (37) that the ratio of resonant frequencies of two isotopes having the same spin is equal to the ratio of their quadrupole moments. This assumes that the internal crystal field of the isotopes is the same (which is true to an accuracy of $10^{-4}$). If the spins of the isotopes are different, the ratio of the quadrupole moments is multiplied by a completely defined and easily computed number. The accuracy of the measurement of the ratio is usually several ten-thousandths.

It can also be seen from expression (37) that nuclear quadrupole resonance permits absolute measurement of the quadrupole interaction constant $\left( eQ \, \dfrac{\partial^2 V}{\partial z^2} \right)$. If the gradient of the electric field is known, the absolute magnitude of the quadrupole moment can be determined. However, in a majority of cases the electric field gradient is only known approximately, so that the accuracy of the absolute determination of quadrupole moments usually is not better than two significant figures.

The nuclear quadrupole moments or, more accurately, the quadrupole interaction constants can also be determined from the NMR energy level shifts which take place due to quadrupole interaction.

Studies of chemical parameters of matter with the help of NMR are mostly based on the "chemical shift" of resonance lines [10, 38]. Chemical shift is caused by the influence of electrons which rotate around the nucleus and exhibit a shielding effect on the nucleus. The magnitude of the shift depends on the type of chemical compound because the electron configuration varies with the form of chemical bond, as well as with the intensity of the applied uniform magnetic field, since the degree of diamagnetic shielding is directly proportional to the intensity of the applied field. The form of chemical bond can be different not only in different chemical compounds but also within one molecule. As an example, let us examine the ethyl alcohol molecule $C_2H_5OH$ (Fig. 37). The electron environment of the hydrogen atoms, which fall in three different groups ($CH_3$, $CH_2$ and OH), is not the same. The magnetic

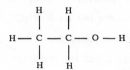

FIG. 37. Configuration of ethyl alcohol molecule

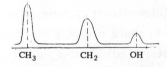

FIG. 38. NMR line shape of protons in ethyl alcohol [10, 68]

resonance of protons of these groups will be observed at slightly different values of the magnetic field. Experiments show that the absorption line has three maxima (Fig. 38), corresponding to the $CH_3$, $CH_2$ and OH groups. The areas under these maxima are in the ratio 3:2:1, corresponding to the number of hydrogen atoms in each group. The line shape shown in the figure was obtained with an applied uniform magnetic field of the order of 550 ka/m; the separation between the peaks of the curves is of the order of 1.6 amp/m. The half-width of the resonance lines corresponds to a change of field of 1/1,000,000 [68]. From these data it becomes clear how homogeneous the field must be so that the instrumental line width will not exceed the characteristic line width due to interaction in the substance. When it is possible to change the field by increments smaller than 1/10,000,000, the ethyl alcohol line shape becomes even more complicated due to appearance of several components in each maximum. This splitting is explained by the effect of the magnetic fields of the protons of one group of atoms on the protons of a second group of atoms in the same molecule, as well as by the interaction of the protons with each other within the same group. In this way, the investigation of nuclear magnetism is an important tool in the hands of chemists in their attempts to establish the structure of complex molecules.

Vast possibilities for obtaining data on chemical structure are also given by the spectra of nuclear quadrupole resonance. It follows from expression (37) that for the case $I = 3/2$ (Cl, B, Br, S) there are only two doubly degenerate quadrupole energy levels. If all similar nuclei assume equivalent locations in the crystal lattice and have identical $Q$, only one absorption line will be observed. If the locations of the atoms containing the investigated nuclei are not chemically or crystallographically equivalent, the quadrupole resonance line will split up, with the number of components corresponding to the number of nonequivalent positions. Thus the crystal lattice of $CCl_4$ has 15 positions of the Cl atoms which differ somewhat structurally, so that the spectrum of crystalline $CCl_4$ consists of 15 lines [38]. By conducting measurements of resonant frequency for the same nucleus in various molecules, one can obtain relative values of the gradient of the internal crystal electric field and make a number of conclusions as to the electronic structure, chemical bonds, etc.*

Measurement of relaxation times can be conducted by several methods [9, 33, 39]. Since the spin-spin relaxation time $\tau_2$ is closely related to the width of the resonance line, $\tau_2$ can be determined from direct measurement of the width of resonance line. The simplest method for measuring the spin-lattice relaxation time $\tau_1$ is as follows. The amplitude of the alternating magnetic field $H_1$ is increased until the absorption signal disappears due to saturation. After this $H_1$ is suddenly decreased to its initial value at which there is no saturation. The absorption signal will appear again and will also increase to its previous value. The increase in signal follows an exponential law with the time constant $\tau_1$, which allows one to determine $\tau_1$.

The most exact and convenient measurements of relaxation times are conducted by the spin-echo method. The essence of one such method for determination of $\tau_2$ is as follows [39]. At a time $t$ after the first 90° pulse a train of 180° pulses with spacings $2t$ is transmitted. Echo signals appear during the

---

*An important application of NMR in investigations of the structure of solids is the so-called Knight shift—a shift in the nuclear resonance frequency in metals due to the presence of conduction electrons. By investigating the change in magnitude of the Knight shift during the transition of a metal from the normal to the superconducting state it is possible to draw conclusions on the number of "paired" electrons. These data are of great importance to the contemporary theory of superconductivity— Editor.

intervals between the 180° pulses (Fig. 39). Due to relaxation processes the amplitude of these signals gradually decreases. The measurement of this decrease of pulse amplitudes as a function of time allows one to determine the spin-spin relaxation time $\tau_2$.

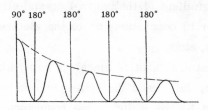

FIG. 39. Determination of relaxation time [33]

Using the method of spin echoes, the spin-lattice relaxation time $\tau_1$ can, for example, be determined as follows. The investigated specimen is irradiated by a sequence of three pulses which respectively rotate the resultant nuclear magnetization vector by 180°, 90° and 180°. After the first pulse the magnetization vector $\vec{M}$ is inverted with respect to the field. Due to the spin-lattice relaxation the vector $\vec{M}$ decreases to zero and then begins to increase again. The effect of the second and third pulses is to rotate the resultant magnetization vector into the plane perpendicular to the direction of the magnetic field, so that a "spreading" fan is created. This is again compressed after the third pulse, causing the spin-echo signal to appear (Section 11). The magnitude of this signal depends on the magnitude of the resultant magnetization vector $\vec{M}$. If the 90° pulse is impressed or "applied" at the time when the vector $\vec{M}$ is equal to zero, the spin-echo signal will not appear. Thus, in this case, the interval between the first and second pulse uniquely determines the longitudinal relaxation time $\tau_1$.

Stabilization and measurement of magnetic fields with the help of NMR have become widely popular and are always used in radio frequency spectroscopic investigations when magnetic resonance is encountered. A magnetic field meter is usually a simple, portable radio frequency spectrometer [13, 29-31]. The absorption signal can be observed conveniently in a known substance, which most often is water with paramagnetic salt impurities [for example, manganese sulfate or iron (III) nitrate]. The line intensity is made sufficient to allow use of small samples and measurement of field strength in small regions. The magnetic field is determined from the value of the resonant frequency of the proton resonance by $H = 18.690 \times 10^{-6} \nu$ ka/m, where $\nu$ is the frequency in c/s. It is seen from this relationship that frequencies in the vicinity of 1-100 Mc/s correspond to field intensities of 15-2000 ka/m. Usually the proton-resonance magnetometers are used for measurement of fields not lower than 10 ka/m. For lower fields the signal voltage will be very small; i.e., the lower limit of applicability is defined by the signal-to-noise ratio. An upper limit does not exist when proton resonance is used. In a field of 1000 ka/m the proton resonance frequency is quite high: 53.5 Mc/s. Therefore, in fields larger than 1000 ka/m a better measurement is obtained when resonance of heavier nuclei is used. In this respect $Li^7$ is very convenient because it has a smaller gyromagnetic ratio than the proton, resulting in a decrease in the utilized frequency to about one third.

Any proton-resonance magnetometer can be used to stabilize the magnetic field, the only required addition being a discriminator. As the absorption line frequency varies due to fluctuations of the magnetic field, an error signal voltage appears at the output of the discriminator. Its sign will be determined by the direction of change of the field. The signal can be amplified and used to regulate the field of the magnet. The simplest method for obtaining the regulating voltage involves use of the derivative of the absorption line, which represents an ideal discriminator curve. By changing the oscillator frequency one can change the magnetic field strength from one stabilized value to the other.

The NMR phenomenon also allows measurements of very low magnetic field intensities. This measurement, based on the effect of nuclear induction, is conducted as follows [10, 11]. A specimen in which NMR can be observed is placed in the field to be measured. A strong magnetizing field $H_m$ (of the order of 10 ka/m or more) is applied perpendicularly to the measured field. The nuclear magnetic moments are aligned along the direction of this field, creating a magnetization $\chi' H_m$. The induction vector $\vec{B}$ in this case is almost completely determined by the component $B_x$. (It is assumed that the measured field direction is along the $z$ axis and the direction of the magnetizing field is along the $x$ axis, so that the $y$ axis is the axis of the receiving coil.) The components $B_y$ and $B_z$, which appear due to precession of magnetic moments, are very small so that the component $B_y$ cannot induce any significant signal in the receiving coil. Then the value of the magnetizing field is rapidly decreased to zero. The magnetic moments start rotating around the direction of the measured field, i.e., around the $z$ axis. Since the magnetic moments were originally oriented along the $x$ axis, their initial rotation will take place in the $xy$ plane, creating measurable components of the induction vector $B_x$ and $B_y$. The change of phase of these components will be determined by the frequency of precession of the magnetic moments around the measured field $H_{meas}$. The resulting component $B_y$ induces an emf in the receiving coil; the rate of change of the emf is recorded and from it the field strength is determined.

## 14. DOUBLE RESONANCE

When nuclei possess quadrupole moments, an important factor determining the line width is the interaction of the quadrupole moment with the internal electric fields. The lattice vibrations in solids perturb the electric potentials at the locations of the nuclei and lead to a quadrupole interaction which changes with time. In liquids the changes in interaction are caused by the molecular motion. In both cases a change in orientation of the quadrupole takes place and, together with this, a change in the orientation of the magnetic moment. The lifetime of the nucleus in a

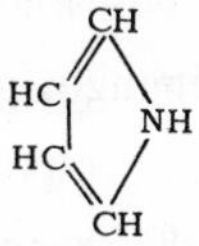

FIG. 40. Structural formula of pyrrole molecule

given state is shortened and the NMR line is broadened. The mechanism of quadrupole relaxation is quite effective, and therefore in a number of cases the broadening of spectral lines is so great that special measures must be taken to detect resonant absorption. As a typical example, let us examine the resonance of pyrrole. The protons themselves do not have a quadrupole moment and it would seem that quadrupole effects should not occur. This, however, is not true at all. From examination of the structural formula of a pyrrole molecule (Fig. 40) it is evident that the protons in the molecule occupy nonequivalent positions: four protons are located near the carbon atoms and one near the nitrogen atom. Due to varying diamagnetic shielding of the nuclei the resonance conditions are somewhat different so that two closely spaced absorption lines are observed. One of these lines is more intense and corresponds to the protons of the CH group; the other, less intense line corresponds to the proton of the NH group. The detection of the second line, however, is not so simple. This is because the nucleus of nitrogen $N^{14}$ has a spin $I = 1$ and its quadrupole moment is not zero. The presence of quandrupole relaxation in the nitrogen nucleus causes a continuous variation in its orientation, and at the same time the orientation of the magnetic moment of the nitrogen nucleus (in whose field the proton of the NH group is located) is also continuously changing. The magnetic coupling of the nitrogen nucleus and the proton also causes the orientation of the proton to change, so that the absorption line for a given proton broadens and is impossible to observe. In order to enable the observation of the magnetic resonance of protons of the NH group it is either necessary to "freeze" the nitrogen nucleus or to increase the continuous variations of the nitrogen nuclear magnetic moment to a level at which the magnetic moment of the proton would detect the magnetic field of the nitrogen nucleus as averaged to zero. This second variation, when the "decoupling of spins" takes place, is easily achieved when the specimen is subjected to an additional radio frequency field, whose frequency coincides with the resonant frequency of nitrogen nuclei (2.9 Mc/s at 750 ka/m) and whose amplitude is sufficient to cause saturation. Now the changes among the magnetic sublevels will become faster than in the case of quadrupole relaxation. Due to this the magnetic quantum numbers of the nitrogen nucleus average effectively to zero and the line width of the protons in the NH group becomes narrower.

The described procedure to increase the influence of magnetic relaxation of nitrogen nuclei is called double resonance [10]. Essentially, double resonance reduces to realization of magnetic resonance simultaneously at two different frequencies.

A second most interesting example of double resonance is the Overhauser effect, in which electron paramagnetic resonance and nuclear magnetic resonance occur simultaneously. It appears that saturation of EPR causes a sharp increase in the intensity of the NMR signal.

The essence of the Overhauser effect is contained in the following [14]. Let us assume that the atoms of the investigated specimen have an electron magnetic moment as well as a nuclear magnetic moment (in the general case these magnetic moments can correspond to different atoms). Let us consider three weakly interacting systems: the nuclear spin system ($N$ system), the electron spin system ($Z$ system) and the lattice ($K$ system). Let us furthermore note that the interaction between the nuclear spin system and the lattice is realized mainly through the electron spin system. If this is so, any relaxational transition which is accompanied by a change of the nuclear magnetic quantum number ($\Delta M = \pm 1$) should, according to the law of conservation of angular momentum, cause a change of the electron magnetic quantum number (by a quantity $\Delta m = \mp 1$). Naturally, the second process—reorientation of the electron spin—can be accompanied by a reorientation of the nuclear spin in the opposite direction.

For every relaxation transition an amount of energy $\Delta E_1 = g\mu_0 H$ is radiated in the electron spin system ($m \to m - 1$). At the same time the transition is accompanied by an absorption of energy $\Delta E_2 = g\mu_{n0} H$ in the nuclear spin system ($M \to M + 1$), which is about $1/1000$ as large as $\Delta E_1$. The remaining energy $\Delta E_3 = \Delta E_1 - \Delta E_2$ is, by conservation of energy, received by the lattice.

Let us denote by $N_M$ the number of particles per unit volume in the nuclear spin system which are in state $M$ and by $T_N$ the temperature of the nuclear spin system. In an analogous way, suppose that $N_m$ and $T_Z$ denote the number of particles per unit volume in the $m$ state and the temperature of the electron spin system, and $N_K$ and $T_K$ denote the same quantities for the lattice. Then $N_{M+1}$, $N_{m-1}$ and $N_{K'}$ will denote the number of particles per unit volume which are respectively in states $M + 1$, $m - 1$ and $K'$. For each pair of interaction particles the probabilities of direct and inverse transitions $M$, $m \rightleftarrows M + 1$, $m - 1$ are the same. Therefore the number of forward transitions will be proportional to $N_m N_M N_K$ and the number of inverse transitions will be proportional to $N_{m-1} N_{M+1} N_{K'}$. In the steady state the number of forward transitions is equal to the number of inverse transitions, i.e., $N_m N_M N_K = N_{m-1} N_{M+1} N_{K'}$. From the Boltzmann distribution, according to which we can write

$$N_m = N_{m-1} e^{-\frac{\Delta E_1}{kT_Z}}; \quad N_M = N_{M+1} e^{+\frac{\Delta E_2}{kT_N}}; \quad N_K = N_{K'} e^{+\frac{\Delta E_3}{kT_K}},$$

we can express the steady-state condition in the form

$$\exp\left[-\frac{\Delta E_1}{kT_Z} + \frac{\Delta E_2}{kT_N} + \frac{\Delta E_3}{kT_K}\right] = 1$$

Using the values of $\Delta E_1$, $\Delta E_2$ and $\Delta E_3$ we obtain

$$\frac{1}{T_N} = \frac{g\mu_0}{g_n \mu_{n0} T_Z} - \frac{g\mu_0 - g_n \mu_{n0}}{g_n \mu_{n0} T_K}$$

The EPR saturation means an increase in temperature of the electron spin system, i.e., $T_Z \to \infty$. Therefore

$$T_N = T_K \frac{(-g_n)\,\mu_{n0}}{g\,\mu_0}$$

Depending upon whether the nuclear $g_n$ factor is positive or negative, the temperature of the nuclear spin system can assume positive or negative values. Let us assume for definiteness that $(-g_n)$ is positive. In this case we conclude that the saturation of the electron resonance causes a decrease in temperature of the nuclear spin system by a factor of $(-g_n)\,\frac{\mu_{n0}}{g\,\mu_0}$, i.e., about 1000. The lowering of the spin system temperature means an increase in population of the lower levels and a decrease in population of the upper levels. At the same time, the excess of the number of nuclei in the lower energy level increases and, correspondingly, the intensity of the NMR signal increases.

At EPR saturation an increase in the number of relaxation transitions from upper to lower levels increases (in the electron spin system). At the same time, as was remarked earlier, a rotation of the nuclear magnetic moments, i.e., reorientations of the nuclei, will take place. Polarization of the nuclei will increase until the number of transitions $m, M \to m - 1, M + 1$ becomes equal to the number of inverse transitions, i.e., until steady state is established.

Experiments with oriented nuclei allow one to obtain a number of valuable data pertaining to the physics of the nucleus. Obviously the polarization of nuclei with the help of double resonance, where complex low-temperature techniques are not required, can greatly simplify the investigation of these problems.

*Chapter V*

**Two-Level Masers**

### 15. BASIC RELATIONSHIPS FOR A TRAVELING-WAVE MASER

"Maser" is an acronym formed from the phrase "microwave amplification by stimulated emission of radiation." At present the term "maser" encompasses a large class of microwave amplifiers and oscillators whose performance is based on the quantum effects of interaction of atoms and molecules with electromagnetic radiation in the microwave region. Such devices are often also called quantum oscillators and amplifiers.

In our discussion of the magnetic resonance phenomenon (EPR and NMR), we saw that in general the effect of resonant interaction of electromagnetic radiation with matter is expressed through absorption of electromagnetic energy. For realization of an amplification (or oscillation) process the number of induced transitions from the upper energy levels to the lower levels must be greater than the number of transitions from the lower levels to the upper levels. For this purpose the energy level population must be reversed (this is referred to as "inversion"). In the simplest case, when the spin is equal to 1/2, this means that a majority of the magnetic moments of the atoms should be brought into a state whose orientation is antiparallel with respect to the applied uniform magnetic field. Description of such nonequilibrium states is facilitated by introduction of the concept of negative spin temperature. The basic relationship here is that which relates populations in two levels (through the Boltzmann factor). It may easily be shown that the temperature $T$ can be expressed in the form

$$T = \frac{E_2 - E_1}{k \ln \dfrac{N_2}{N_1}} \tag{40}$$

where $N_1$ is the number of spins in the lower level and $N_2$ is the number of spins in the upper level.

The numerator of this expression is always positive, since $E_2 > E_1$. In thermal equilibrium $N_2 < N_1$ and the natural logarithm is negative, making $T > 0$. At saturation, when $N_1 = N_2$, the logarithm is zero and $T = \infty$. For an inversion of energy levels, when $N_2 > N_1$, the logarithm is positive and the temperature negative.

Quantum paramagnetic amplifiers are based on the same physical processes that characterize the phenomenon of magnetic resonance. Such concepts as spin system and lattice, transverse and longitudinal relaxation times, spontaneous and induced transitions, saturation, etc., still apply.

Our theoretical description of the performance of two-level masers will be in terms of the model studied in Chapter II. In particular, Eq. (22) (Section 4) for the average absorbed power will, in the case of a paramagnetic amplifier, express the average radiated power because the system is now in a negative temperature state. It must, of course, be assumed that the relaxation mechanisms which tend to return the system to its equilibrium state have little effect; i.e., the relaxation transition of the paramagnetic substance toward equilibrium is relatively slow (compared with the processes caused by the stimulating field).

The simplest amplifier can be imagined as a piece of waveguide filled with a paramagnetic substance in its active state, i.e., in a state when $n_{21} = N_2 - N_1 > 0$. A traveling wave proceeding down the amplifier will cause induced emission and an exponential increase in power so that the power gain can be expressed as $G = \exp(al)$, where $l$ is the length of the waveguide and $a$ is the molecular gain coefficient. Simultaneously with the amplification in the waveguide there are also energy losses caused by the finite conductivity of the waveguide walls and the presence of dielectric losses in the paramagnetic substance. Suppose that the general attenuation coefficient for the traveling wave is $a_g$. Then the condition for amplification in a traveling-wave amplifier can be expressed as $a > a_g$. Obviously it is desirable that the gain coefficient be much larger than the attenuation coefficient. In order to accomplish this it is first necessary to assure the absence of saturation and to maintain a sufficiently large excess of atoms in their radiating state. From Eq. (22) it follows that the condition that prevents saturation is

$$\gamma^2 H_1^2 \tau_1 \tau_2 \ll 1 + (\omega_0 - \omega)^2 \tau_2^2 \tag{41}$$

When this condition is satisfied the quantum amplifier should work in the linear regime. For the case $\omega = \omega_0$ the linearity condition becomes

$$H_1 \ll \frac{h}{2\pi g\mu_0} \sqrt{\frac{\tau_2}{\tau_1}} \Delta\omega \tag{42}$$

where we make use of the fact that $\gamma = 2\pi g\mu_0/h$ and that $\tau_2^{-1} = \Delta\omega_2$. Assuming that the molecular line width $\Delta\omega_2 \sim 10^6 \ \text{sec}^{-1}$, $g \sim 2$ and letting $\tau_2 \approx \tau_1$, we get $H_1 \ll 4 \ \text{amp/m}$. Thus the input power level should be significantly smaller than 1 watt (for a waveguide cross-sectional area of $\sim 1 \ \text{cm}^2$). The obtained estimate also indicates that one should not expect high power levels in quantum amplifiers.

Using Eqs. (22) and (41), we write an expression for the total radiated power at resonance ($\omega = \omega_0$):

$$P_{\text{tot}} = \frac{2N\mu^2 \omega_0^2 \tau_2 Sl}{\mu' kT_s} \cdot \frac{\mu' H_1^2}{2} \tag{43}$$

where $S$ is the cross-sectional area of the waveguide and $NSl$ is the total number of spins.

The quantity $\frac{1}{2}\mu' H_1^2$, representing the energy density of the incident radiation ($\mu'$ is the magnetic permeability), is related to the input power $P_{\text{in}}$ by the expression

$$P_{\text{in}} = \frac{1}{2}\mu' H_1^2 v_g S \tag{44}$$

where $v_g$ is the group velocity of the wave propagated in the waveguide. Taking into account this last relationship we get

$$P_{\text{tot}} = \frac{2N\mu^2 \omega_0^2 \tau_2 l}{\mu' kT_s v_g} P_{\text{in}} \tag{45}$$

On the other hand, the power radiated by the molecules in the volume $S dz$, where $z$ is taken along the direction of propagation, is equal to

$$dP = P_{in} a\, dz$$

from which

$$a = \frac{dP}{dz} \bigg/ P_{in}$$

If the radiation is uniform

$$\frac{dP}{dz} = \frac{P_{tot}}{l} \qquad \text{and} \qquad a = \frac{P_{tot}}{l P_{in}} \tag{46}$$

Using now expression (45) we get

$$a = \frac{2 N \mu^2 \omega_0^2 \tau_2}{\mu' k T_s v_g} \tag{47}$$

Since $N = N_1 + N_2$ and $N_1 = N_2 \exp\left(-\dfrac{2\mu H}{k T_s}\right)$, we have $\dfrac{N\mu H}{k T_s} = n_{12}$

or $\dfrac{N\mu}{k T_s} = 2\pi n_{12} g \mu_0 / h \omega_0$. Therefore the molecular gain coefficient can also be written as

$$a = \frac{4\pi g \mu_0 \mu \omega_0 \tau_2 n_{12}}{\mu' h v_g} \tag{48}$$

Let us now make some remarks concerning the derived formulas. First we note that the total radiated power $P_{tot}$ and the molecular gain coefficient $a$ are negative quantities. Actually, in expressions which define $P_{tot}$ and $a$ all quantities are positive except $T_s$ and $n_{12}$. These last two quantities are negative because in our case the system is in the radiating state so that $N_2 > N_1$, i.e., $N_1 - N_2 = n_{12} < 0$ and $T_s < 0$. There is nothing strange about the fact that $P_{tot}$ and $a$ are negative, however. In the derivation of the expressions which define $P_{tot}$ and $a$ we have used at the start the expression for the average power of magnetic resonance absorption so that the presence of a negative sign is only a mathematical expression of the fact that with $N_2 > N_1$ the "absorption" is negative; i.e., radiation actually takes place.

The physical meaning of the above formulas is fully understood and is as follows. The molecular radiation coefficient and the total radiated power increase with an increase in radiation frequency $\omega_0$. This is because with an increase in $\omega_0$ the magnitude of the quantum $\dfrac{h}{2\pi}\omega_0$ transferred to the amplified wave by every radiating atom also increases. With an increase in the magnetic moment $\mu$, the coefficient $a$ and the power $P_{tot}$ also increase. This takes place because with an increase in $\mu$ the coupling between the amplified wave and the radiating spin system increases. Also obvious is the dependence of $a$ on $n_{21} = -n_{12}$: the greater the excess of radiating molecules, the greater will be the amplification. Also let us note that the greater the excess of atoms in the upper energy level, i.e., the greater $N_2/N_1$, the smaller will be the absolute value of $T_s$ and the greater the amplification. The amplification also increases when $\tau_2$ is increased. The time $\tau_2$ describes the time interval during which the atom interacts with the radiation field before undergoing a transition. Consequently, the larger $\tau_2$, the larger will be the probability that the atom will give up its energy in the form of a photon and that it will participate in the amplification process.

An interesting fact is that the radiated power $P_{tot}$ and the molecular gain coefficient depend on the velocity $v_g$ (the smaller

$v_g$, the larger the gain). This dependence shows that utilization of slow-wave structures will significantly decrease the length of amplifier required for a given amount of amplification. In fact, the use of slow-wave systems increases the interaction time between the electromagnetic wave and the substance. In this sense the retardation of the wave is, without doubt, preferable to the distribution of the radiating substance over a longer interval. There is, of course, one more possibility, in which a significant electrical length can be combined with reasonable geometric dimensions, using a resonant cavity. We will return later to the problem of masers of the resonant cavity type.

Let us now numerically estimate the amplification coefficient. Letting $N = 2 \times 10^{17}$ spins/cm$^3$, $\omega = 2\pi\nu = 6.28 \times 10^{10}$ sec$^{-1}$, $T_s = -1°$K, $\tau_2 = 10^{-7}$ sec, $\mu' \cong 4\pi \times 10^{-7}$ henry/m,* $k = 1.38 \times 10^{-23}$ joule/deg, $\mu \approx 10^{-29}$ joule·m/amp and $v_g = 3 \times 10^{10}$ cm/sec, we get a molecular gain coefficient of the order of $a \approx 0.005$ cm$^{-1}$. In order to obtain a power gain $G = \exp(al)$ of the order of 200 (23 db) a waveguide maser about 10 m long ($al \sim 5$) is necessary. The construction of such a maser is hardly justified. However, if the molecular gain coefficient is computed for a slow-wave structure with $v_g \approx 0.01\ c$, we get $a \approx 0.5$ cm$^{-1}$ and for the same gain of 23 db the required length of the maser will be only 10 cm.

Let us now investigate the bandwidth of a traveling-wave quantum amplifier. The bandwidth is here defined as the frequency interval between the points at which the power gain is one half of the value at the center frequency. From

*Permeability of vacuum in the rationalized MKS and corresponding systems.

expression (46) and taking into account (22), (47) and (41) we get

$$a = \frac{C}{(\omega_0 - \omega)^2 + \left(\dfrac{1}{\tau_2}\right)^2} \tag{49}$$

where $C = N\mu^2 \omega_0^2 H_1^2 / \mu' k T_s v_g \tau_2$ and is independent of frequency. The power gain is

$$G = \exp\left[\frac{lC}{(\omega_0 - \omega)^2 + (\Delta\omega)^2}\right] \tag{50}$$

By definition, the bandwidth will be determined from the relationship

$$\exp\left[\frac{lC}{B^2 + (\Delta\omega)^2}\right] = \frac{1}{2} G(0)$$

from which

$$B = \left[\ln 2 \middle/ \ln \frac{1}{2} G\right]^{\frac{1}{2}} \Delta\omega \tag{51}$$

where $G(0) = G$ is the gain at the center frequency. Since usually $\ln \frac{1}{2} G \gg \ln 2$, the bandwidth of a maser will always be smaller than the width of the magnetic resonance line. In particular, returning to our previous numerical example and letting $G \simeq 200$, we get $B \simeq 0.4 \Delta\omega$.

The main method for increasing the bandwidth is to increase the molecular bandwidth $\Delta\omega$. But an increase in $\Delta\omega$ is equivalent to a decrease in $\tau_2$, and a decrease in $\tau$ will cause a decrease in gain, as is evident from Eq. (48). Consequently, if it is desired to preserve a certain gain and increase the bandwidth at the same time it becomes necessary to change the frequency or the magnetic moment or the concentration of spins in order to compensate for the decrease in $\tau_2$. Satisfying all these mutually exclusive requirements is one of the main difficulties encountered in maser construction. Let us also note that the bandwidth of a maser is wider when solid paramagnetic substances are used instead of gases. This follows from the fact that in solid paramagnetic substances a higher spin concentration and a wider molecular bandwidth can be achieved. Generally the bandwidth of a solid-state paramagnetic amplifier is several tens of megacycles.

Let us now investigate the efficiency of a traveling-wave quantum amplifier. In order to maintain the excess population of excited atoms $n_{21}$, each atom must be supplied at least one quantum of energy $\dfrac{h}{2\pi}\omega_0$ during the time interval $\tau_1$. Therefore the amplifier must be supplied with power of the order of

$$P_s \simeq n_{21}\frac{h\omega_0}{2\pi\tau_1}$$

On the other hand, according to (43), the average radiated power per unit volume can be written as

$$P \approx \frac{2\pi n_{21}g\mu_0^2\omega\tau_2 H_1^2}{h} \tag{52}$$

Defining the efficiency $\eta$ as the ratio of the power radiated by the atoms to the input power, we find that

$$\eta \approx \frac{4\pi^2 g\mu_0^2\tau_1\tau_2 H_1^2}{h^2} \tag{53}$$

Remembering now our linearity condition and letting $g \approx 2$ we get $\eta \ll \dfrac{1}{2}$. If we now take into account the energy losses in the electronic apparatus associated with the maser, the efficiency will be even smaller (significantly less than 1%).

Finally, let us investigate the noise figure of a traveling-wave quantum amplifier. By definition, the noise figure $F$ of any amplifier is the ratio of the total noise power at the amplifier output to the output noise power of purely thermal origin. The appearance of internal noise in a maser is caused by thermal radiation from the waveguide walls and by noncoherent spontaneous radiation of excited molecules.

A detailed consideration of all factors which contribute to the noise figure of the waveguide type of quantum amplifier leads to the expression [41, 42]

$$F = \left(1 + \frac{a_g}{a - a_g}\right)\left(1 + \frac{N_2}{N_2 - N_1} \frac{h\nu}{P_p}\right) \tag{54}$$

where $P_p$ is the thermal noise power per unit frequency interval and at temperature $T$ (for one oscillator).

The power $P_p$ is determined from Planck's formula

$$P_p = \frac{h\nu}{e^{\frac{h\nu}{kT}} - 1} \tag{55}$$

The noise figure $F$ will be small when the waveguide attenuation coefficient $a_g$ is small and when the number of spins $N_2$ in the upper energy level is large compared to the population in the lower level $N_1$. Considering that in the centimeter wavelength region and below $h\nu \ll kT$ for temperatures of almost $4°K$ and expressing $N_2/(N_1 - N_2) \approx kT_s/h\nu$ we get for the case $a \gg a_g$

$$F = 1 - \frac{T_s}{T_0} \tag{56}$$

where $T_0$ is room temperature and $T_s$ is the spin-system temperature.

In this expression the temperature $T_s$ is negative so that $F > 1$ at all times. With full inversion (when the populations of the levels have changed places, so to speak) the absolute value of this temperature does not change; only its sign changes. If before the inversion the spin system was in equilibrium with the lattice, then $-T_s$ should have the meaning of the temperature of the paramagnetic substance. Therefore a decrease in the operating temperature of the maser leads to a significant decrease in the noise figure and a large increase in the sensitivity of the maser. The noise figure can reach values of the order of 1.004 (0.02 db) if the temperature is of the order of 1.2°K. This is a very low noise figure, which cannot be achieved in practice with ordinary amplifiers. This property alone is the most valuable characteristic of a quantum amplifier and is the reason for its intensive development.

The noise figure of a quantum amplifier is very often expressed in terms of the noise temperature $T_M$ [40]. The noise temperature $T_M$ is related to the noise figure $F$ by the relationship

$$T_M = (F - 1)\,290°K \tag{57}$$

If, for example, the noise figure of a maser is 1.004 (0.02 db), its noise temperature $T_M$ is equal to 1.16°K. In real masers the noise temperature is somewhat higher, usually of the order of 10°K. For comparison let us note that the noise temperature of a good radar receiver in the 10-cm wavelength region is about 1500-2500°K [43].

## 16. RESONANT-CAVITY QUANTUM AMPLIFIERS AND OSCILLATORS

Besides wave-retardation systems, resonant cavities which also assure a sufficiently long electrical path with a short geometrical length can be utilized in maser construction. Suppose that a cavity contains a paramagnetic substance with a negative spin temperature and that the transition frequency is the same as the resonant frequency of the cavity. In this case the electromagnetic field contained in the resonant cavity will cause induced radiation of the paramagnetic system. If the radiation exceeds the losses in the cavity walls, amplification of microwave power will take place. Since a resonant cavity is a system with positive feedback, oscillation is also possible. For this, however, it is necessary for the radiated power to exceed not only the losses in the resonant cavity walls but also the losses in external circuits. The conditions for gain and oscillation of a cavity type maser can be expressed mathematically by the use of the figure of merit ($Q$) concept.

According to a general definition the (loaded) $Q_L$ of a resonant cavity is

$$Q_L = \omega \frac{W}{P_L} \tag{58}$$

where $W$ is the energy stored in the resonant cavity and $P_L$ is the total power loss.

The total losses consist of dielectric losses and losses in the cavity walls $P_0$, "magnetic losses" in the working substance $P_m$, and radiation losses $P_e$. Thus we can write

$$\frac{1}{Q_L} = \frac{P_0}{\omega W} + \frac{P_m}{\omega W} + \frac{P_e}{\omega W}$$

or

$$Q_L^{-1} = Q_0^{-1} + Q_m^{-1} + Q_e^{-1} \tag{59}$$

Let us note that magnetic losses in a system with negative temperature actually represent radiation of energy. Therefore the quantity $P_m$ and the magnetic figure of merit $Q_m$ for such systems will be negative. Using the above introduced notation we can write the condition for oscillation in the form

$$-Q_m^{-1} > Q_0^{-1} + Q_e^{-1} \tag{60}$$

Analogously, the condition for amplification has the form

$$Q_0^{-1} + Q_e^{-1} > -Q_m^{-1} > Q_0^{-1} \tag{61}$$

Remembering Eq. (22) (Section 4) for the average radiated power and using relationship (41) we can define the "magnetic losses" $P_m$ as

$$P_m = \int_{V_m} P\,dV = \frac{N\mu^2\omega_0^2\tau_2}{kT_s} \int_{V_m} H_1^2\,dV$$

where $V_m$ is the volume occupied by the paramagnetic substance.

The energy $W$ stored in the resonant cavity is given by the relationship

$$W = \frac{\mu_0'}{2} \int_V H_1^2\,dV$$

where $V$ is the cavity volume and $\mu_0'$ is the magnetic permeability of vacuum.

With these expressions we compute $Q_m$ as

$$Q_m = \frac{\mu_0' kT_s}{2N\mu^2\omega_0\tau_2\xi} \tag{62}$$

where

$$\xi = \frac{\displaystyle\int_{V_m} H_1^2 \, dV}{\displaystyle\int_V H_1^2 \, dV} \tag{63}$$

and is the filling factor of the resonant cavity.

In order to satisfy the conditions for oscillation and amplification it is desirable that the magnetic figure of merit $|Q_m|$ be as small as possible. A small $|Q_m|$ means a large radiated power $|P_m|$. A small $|Q_m|$ may be achieved by an increase of the number of spins per unit volume, $N$, the spin–spin relaxation time $\tau_2$ and the cavity filling factor $\xi$, as well as a higher degree of inversion, which yields lower values of $|T_s|$. An increase in $\tau_2$ can be obtained by diamagnetic dilution, but this will cause a decrease in the spin density $N$. A compromise solution usually is a spin density of the order of $10^{18}$ spins/cm$^3$. The limit of increase of the filling factor $\xi$ is determined by the dielectric losses in the paramagnetic substance. When the dielectric losses are increased, $Q_0$ decreases. Thus the general requirements may be reduced to having a substance with narrow lines (large $\tau_2$) and small dielectric losses, making the filling factor as large as possible and the temperature as low as possible (the lower the temperature $T$, the easier it is to obtain small $|T_s|$). Let us finally also remark that low $Q_m$ is obtained more easily at higher frequencies.

The above conditions for oscillation and amplification can be expressed in terms of the maser gain $G$. The main prerequisites for determination of $G$ appear to differ depending on how the resonant cavity is coupled to the transmission line: whether the cavity is a transmission cavity or a reflection cavity. In the former case the resonant cavity is connected to two waveguides, one at the input and the other at the output. In order to protect the maser from load reflections and to prevent the radiation from entering the input waveguide, ferrite isolators are usually used, as shown in Fig. 41. In order to be able to

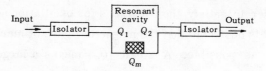

FIG. 41. Transmission–cavity maser

compare the output and input powers it is necessary to determine the cavity transmission loss. For this (Fig. 42), the resonant cavity is connected to only one waveguide and the separation of powers (incident on the cavity and propagated from the cavity) is accomplished by some kind of directional system (e.g., a ferrite circulator). The reflection coefficient can be used in this case to compare the input and output powers.

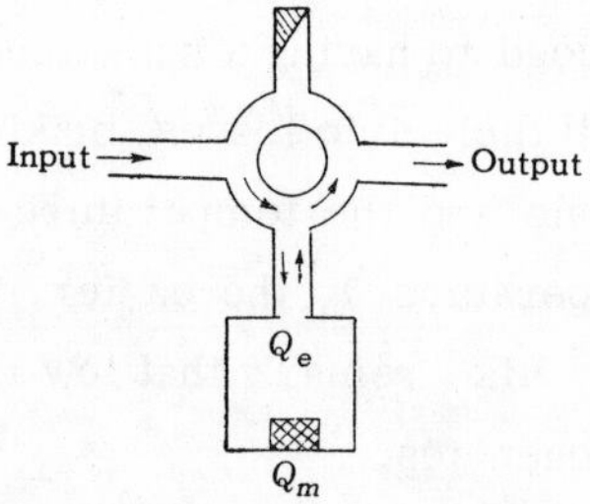

FIG. 42. Reflection–cavity maser

Using the transmission line theory it can be shown that for a transmission-cavity maser the ratio of the output power to the input power is given by the expression

$$G = \frac{4Q_1^{-1}Q_2^{-1}}{\left(Q_0^{-1} + Q_m^{-1} + Q_1^{-1} + Q_2^{-1}\right)^2} = \frac{4Q_L^2}{Q_1Q_2} \tag{64}$$

where $Q_L^{-1} = Q_0^{-1} + Q_m^{-1} + Q_1^{-1} + Q_2^{-1}$, and $Q_1$ and $Q_2$ are coupling $Q$'s. It follows from the relationship obtained for $G$ that if there is no paramagnetic substance in the cavity and if the cavity itself is ideal ($Q_0^{-1} = 0$) then $G \geq 1$ (the equal sign holds for $Q_1 = Q_2$). Actually we always have $Q_0^{-1} > 0$ so that always $G < 1$ and no amplification is possible until $|Q_m^{-1}| < Q_0^{-1}$. Consequently, it is required for amplification that $Q_m^{-1}$ be negative and that its absolute value be greater than $Q_0^{-1}$, i.e., $|Q_m^{-1}| > Q_0^{-1}$. If at the same time $|Q_m^{-1}|$ reaches the value of $Q_0^{-1} + Q_1^{-1} + Q_2^{-1}$, we have $G = \infty$; i.e., the system becomes oscillatory. Thus we have obtained the same conditions for amplification and oscillation as were derived at the beginning of this section. The amplification will be maximum if a) the figure of merit of the resonant cavity, $Q_0$, is large in comparison with $|Q_m|$, $Q_1$ and $Q_2$; b) the figures of merit $Q_1$ and $Q_2$ are nearly equal $Q_1 \simeq Q_2 \simeq Q_e$; and c) $Q_m^{-1}$ is nearly equal to $Q_1^{-1} + Q_2^{-1} = 2Q_e^{-1}$. With these assumptions the expression for the gain $G$ becomes

$$G = \frac{\left(2Q_e^{-1}\right)^2}{\left(2Q_e^{-1} - |Q_m^{-1}|\right)^2} \tag{65}$$

The resonant line width of the paramagnetic substance utilized in a maser is generally larger than the width of the resonance curve of the loaded cavity (it is, of course, assumed that the figure of merit $Q_L$ is sufficiently large). In this case the amplification bandwidth of the maser $B$ will be determined by the bandwidth of the resonant cavity, i.e.,

$$B = \nu_0 Q_L^{-1} \simeq \nu_0 \left(2Q_e^{-1} - |Q_m^{-1}|\right) \tag{66}$$

It is clear from this that the amplification bandwidth $B$ and the gain $G$ are related by the expression

$$B\sqrt{G} = \nu_0 2Q_e^{-1}$$

Considering that $|Q_m^{-1}| \simeq 2Q_e^{-1}$, we get

$$B\sqrt{G} \simeq \nu_0 |Q_m^{-1}| = \text{const} \tag{67}$$

Therefore an increase in gain causes a decrease in bandwidth which decreases as the square root of the gain. Also, vice versa, an increase of the bandwidth $B$ causes the gain to drop. The magnitude of this product depends on the magnetic figure of merit $Q_m$: the smaller $Q_m$ the larger $B\sqrt{G}$. Methods for decreasing $|Q_m|$ were analyzed by us previously.

For a maser with a reflection resonant cavity the gain, which actually represents the reflection coefficient, has the form

$$G \approx \left(\frac{Q_e^{-1} + |Q_m^{-1}|}{Q_e^{-1} - |Q_m^{-1}|}\right)^2 \tag{68}$$

In deriving this expression $Q_0^{-1}$ was neglected compared with $Q_m^{-1}$ and $Q_e^{-1}$, and a negative $Q_m$ was assumed. The gain will be maximum when $|Q_m|$ is nearly equal to $Q_e$. Under this condition the expression for the gain becomes

$$G \simeq \frac{\left(2Q_e^{-1}\right)^2}{\left(Q_e^{-1} - |Q_m^{-1}|\right)^2} \tag{69}$$

It is clear from this that the expression for $G$ in the case of a reflection cavity is very similar to the expression for the gain $G$ of a maser with a transmission cavity [Eq. (65)]. The difference lies in the absence of the coefficient 2 in front of the first term in the denominator. This means that the gain of a reflection maser is larger than the gain of a transmission cavity maser. The reason for this is that in a transmission cavity the molecules radiate power into the input waveguide as well as into the output waveguide, while in a reflection cavity there is only one output. With $\left(Q_m^{-1}\right)_{\mathrm{refl}} = \left(Q_m^{-1}\right)_{\mathrm{trans}}$ and $\left(Q_e^{-1}\right)_{\mathrm{refl}} = Q_1^{-1} + Q_2^{-1} \simeq 2\left(Q_e^{-1}\right)_{\mathrm{trans}}$ we get $G_{\mathrm{refl}} = 4 G_{\mathrm{trans}}$. It is obvious from this that the reflection maser is more effective, even though it may be less convenient.

By analogy to the transmission cavity case we can write the expression for the bandwidth of a reflection maser in the form

$$B = \nu_0 Q_L^{-1} \simeq Q_e^{-1} - |Q_m^{-1}| \tag{70}$$

The relationship between the gain and the bandwidth $B$ is analogous to Eq. (67). But because now we have $|Q_m^{-1}| \simeq Q_e^{-1}$, we get

$$B\sqrt{G} = 2\nu_0|Q_m^{-1}| \tag{71}$$

It is clear from this that with identical gains and identical $Q_m$ a maser with a reflection resonant cavity has twice as large a bandwidth (or twice the gain for the same bandwidth).

In conclusion, let us note that for high gain, when $|Q_m| \sim Q_e$, small relative changes in $Q_m$ and $Q_e$ will strongly affect the magnitude of the gain and the bandwidth. This statement pertains to masers with a transmission cavity as well as those with a reflection cavity. We shall see later that this instability is a peculiar property of two-level masers. Small variations in $Q_m$ and $Q_e$ are less important in three-level resonant cavity masers.

## 17. METHODS FOR ACHIEVING INVERSION
### [14, 40, 42]

Up to this point we have formally examined the performance of a maser assuming that somehow a negative spin temperature

$T_s$ was obtained in an active substance; i.e., an inverted population of the energy levels was present. Let us now examine the methods used to achieve this inversion.

*180° pulse inversion.* This method of inverting the population of the Zeeman energy levels has much in common with the spin-echo method (Section 6). Under the influence of a pulse of an RF magnetic field at the resonant frequency and of length $t_1$, such that $\gamma H_1 t_1 = \pi$, the resultant magnetic moment will be rotated by 180°, as a result of which the system will be put into a state with negative spin temperature. Considering that $\gamma = 2\pi g \mu_0 / h$ and assuming $g \approx 2$, we obtain a value for the product $H_1 t_1$ of the order of $1.5 \times 10^{-5}$ amp·sec/m.

It is seen that in order to obtain complete inversion, two conditions must be rigorously satisfied. First, the frequency of the pulse must be equal to the transition frequency. Second, the duration and intensity of the pulse must be accurately determined in order to satisfy the previous requirement for the product of the two quantities. Furthermore, the inversion must be accomplished in a time that is much shorter than the spin-lattice relaxation time. If the length of the pulse is comparable to or exceeds the time $\tau_1$, the spins will return to their initial state before the inversion process is completed. But since for the sample materials usually used in two-level masers the spin-lattice relaxation time at low temperatures lies in the range from milliseconds to several seconds, this last requirement is not a serious handicap. A more drastic limitation is imposed by the magnitude of $H_1$, which should exceed the width of the resonance line $\Delta H$. The width of the line is essentially determined by the spin-spin interaction and the magnitude of the local internal random magnetic fields. If the local fields exceed or are comparable to the alternating field $H_1$, they will prevent full inversion from taking place. Consequently, the magnitude of the field $H_1$ should be appreciably greater than the width of the resonant line $\Delta H$. The width of the line $\Delta H$ can be of the order of a fraction of an amp/m or higher. The corresponding increase in $H_1$ can lead to completely unacceptable values for the length of the pulse $t_1$. For example, if $\Delta H$ is of the order of 1 ka/m, it is required that $H_1$ be at least of the order of 5-10 ka/m, with the result that $t_1$ will be of the order of $1.5$-$3 \times 10^{-9}$ sec.

In the case of resonant-cavity masers a narrow pulse width imposes some definite requirements on the cavity $Q$. If the $Q$ is high, the cavity excited by the pulse will "ring," and this will decrease the degree of inversion. It follows from this that in order to obtain a maximally complete inversion it is necessary to have a resonant cavity with a low figure of merit ($Q \sim 100$). This requirement, however, contradicts the fact that in order to have a maser working in an amplification mode a resonant cavity with as high a $Q$ as possible is necessary. Several methods were proposed to satisfy these contradictory requirements. For example, it is possible to obtain the inversion in one resonant cavity (with low $Q$) and then transfer the paramagnetic substance to a high-$Q$ cavity in which the amplification takes place; it is possible to "switch" the $Q$ of a resonant cavity, making it low during the time of inversion (e.g., by spark discharge, gap breakdown, magnetostrictive deformation, etc.); it is also possible to use a resonant cavity that is designed for two types of oscillations: for one type of oscillation the cavity has a low $Q$ (inversion) and for the second it has a high $Q$ (amplification and oscillation), etc.

*Inversion by adiabatic fast passage.* An RF magnetic field $H_1$ whose frequency differs widely from the transition frequency has practically no effect on the precession of the magnetic moments. As was pointed out before, at resonance in a rotating coordinate system a precession of magnetic moments around the vector $H_1$, which is perpendicular to the direction of the dc magnetic field, will take place. Obviously, intermediate cases are also possible. Figure 43 shows the behavior of spins during the passage of the RF magnetic field through the resonant frequency. At any instant there exists some effective magnetic field $H_{eff}^*$ around which the precession of the magnetic moments takes place. In the laboratory coordinate system the tip or point of the precessing vector $\vec{M}$ will describe a spiral on a sphere of radius $\vec{M}$. As the RF magnetic field passes through resonance, the direction of the resultant magnetic moment changes by 180°; i.e., the population of the energy levels is reversed.

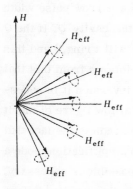

FIG. 43. Several stages of spin inversion with adiabatic fast passage (in the rotating coordinate system) [40]

Let us examine the conditions that should be satisfied during inversion by adiabatic fast passage. First, the passage should be fast in the sense that the total time of the passage $\Delta t$ should be significantly smaller than the spin-lattice relaxation time $\tau_1$. If this is not so the competing relaxation processes will not allow the

---

*In the rotating coordinate system the effective field $\vec{H}_{eff} = \vec{H}_{rot} + \vec{H}_1$, where $\vec{H}_{rot} = \vec{H} + \frac{\vec{\omega}}{\gamma}$ ($\vec{\omega}$ is the angular velocity vector representing the rotation of the coordinate system). At resonance the frequency of the coordinate system rotation coincides with the precession frequency $\omega = \gamma H$, so that $H_{rot} = 0$ and $H_{eff} = H_1$.

establishment of an excess population in the upper energy level. Since $\tau_1$ usually ranges from fractions of a millisecond to several seconds we should have $\Delta t \ll 10^{-6}$ sec. On the other hand the passage should be sufficiently slow to allow the molecular system to follow the changes in $H_{eff}$. This is where the "adiabatic" character of the passage appears. If the passage speed is high, the system cannot adiabatically follow all these changes, and transient states which are analogous to those that determine the pulse inversion will be induced. The parameter that can characterize the duration of internal motion can be the quantity $\omega_{eff}^{-1}$, where $\omega_{eff}$ is the angular frequency of precession of magnetic moments around $H_{eff}$ (i.e., $\omega_{eff} = \gamma H_{eff}$). Consequently, the condition for the adiabatic nature of the passage will be that the relative change in $H_{eff}$ during the time $1/\omega_{eff}$ should be negligible. From this we obtain that $\Delta t \gg 1/\omega_{eff}$. Since the smallest value of $H_{eff}$ is $H_1$ and since $\omega_{eff} = \gamma H_{eff} > \gamma H_1 = \omega_1$, it follows that $\Delta t \gg (\gamma H_1)^{-1}$. In order to estimate $\Delta t$ numerically it is necessary to assume a value for $H_1$. The magnitude of $H_1$ should be appreciably larger than that of the internal local fields (otherwise there will be no precession of spins around the effective field $H_{eff}$). For a line width $\Delta H$ of the order of 100 amp/m, the field $H_1$ should be at least of the order of 1 ka/m. Since $\gamma = 2\pi g \mu_0/h \cong 2 \times 10^5$, we get $\Delta t \gg 0.5 \times 10^{-8}$ sec. Thus we see that the condition of adiabaticity ("slowness" of passage) does not contradict the condition of a relatively fast passage because the time for the passage appears to lie in the interval from $10^{-6}$ to $0.5 \times 10^{-8}$ sec. If the passage of the EPR line is accomplished by changing the magnitude of the magnetic field instead of changing the frequency, then for a line width of 100 amp/m the rate of change of the field is $\sim 10^9$ amp/m·sec.

This, of course, is very approximate. If, for example, $\tau_1 = 1$ sec and $H_1 \sim 1$ ka/m, a rate of change of the field of $10^4$ amp per m·sec is fully acceptable. In any case the above estimates show that inversion by an adiabatic fast passage is technically completely feasible.

Finally, let us mention one more important fact. During the reorientation of the resultant magnetic moment $\vec{M}$ an emf is induced in the cavity. According to Lenz's law it will oppose the cause by which it was induced—the reorientation of the magnetic moment—and will oppose the inversion of the energy levels. Therefore a sufficiently high energy density of the exciting field is required for a successful inversion. From this point of view the inversion by an adiabatic fast passage is less convenient compared with the pulse inversion because it requires higher power levels of the RF magnetic field. (For pulse inversion the induced emf is absent and the inversion is completed before the induced voltage "builds up.") However, the inversion by an adiabatic fast passage has a number of advantages compared with the pulse inversion method. The rigorous requirement that the frequency of the alternating magnetic field must be exactly equal to the transition frequency is removed, it is not necessary to control accurately the duration of application of the exciting field, the passage time $\Delta t$ can vary over a relatively large interval, and the magnitude of the RF magnetic field becomes independent of the interaction time. It is also obvious that the direction of the frequency change during the passage through resonance is immaterial and that the passage through resonance can be accomplished either by changing the frequency of the RF field or by varying the magnitude of the static magnetic field.

*Inversion by sudden field reversal* consists in a quick non-adiabatic change of the sign of the applied static field *H*. The change in polarity of the field should be so fast that it does not allow the spin system to react to this change. In this case the resultant magnetic moment after inversion of the field *H* will be directed opposite to the field, the former lower state with large population will become the upper state, and the former upper state will become the lower state. This condition corresponds to a negative spin temperature, and until the relaxation processes establish an equilibrium the spin system will be in a state that is capable of radiation, i.e., amplification and oscillation.

Now let us investigate how fast the polarity of the static field should be changed. The spin system will not be able to react to the change of the field if this change is faster than any atomic motion. In particular, the reversal of the field should occur during a time $\Delta t$ that is appreciably smaller than the period of precession of the spins around the static field *H*. Thus $\Delta t \ll 2\pi/\omega$, where $\omega = \gamma H$. In fields of the order of 250 ka/m we have $\omega$ of the order of $6 \times 10^{10}$ sec$^{-1}$ ($\gamma = 2\pi g\mu_0/h \simeq 2 \times 10^5$ m/a·sec) so that $\Delta t \ll 10^{-10}$ sec. This estimate imposes stringent requirements on the rate of change of the static field. It is evident that the field should change from +250 ka/m to -250 ka/m at a rate greater than $5 \times 10^{15}$ amp/m·sec. The required rate of change of the field can be substantially decreased, however, if one proceeds as follows: the static field *H* is adiabatically slowly decreased to some minimum value $H_{min}$, then the polarity of the field $H_{min}$ is suddenly changed, and finally the field is slowly adiabatically changed from $-H_{min}$ to the value $-H$. The smaller the value of $H_{min}$, the smaller will be the precession frequency of the magnetic moments and the easier it will be to accomplish a sudden refersal of the field. The minimum allowed value of the static field is determined by internal local magnetic fields. Therefore $H_{min}$ cannot be arbitrarily small and should always be considerably larger than the width $\Delta H$ of the resonance line. If, for example, $\Delta H \sim 10$ amp/m then taking $H_{min} \sim 100$ amp/m we obtain $\omega \simeq 2 \times 10^7$ sec$^{-1}$ and $\Delta t \ll 3 \times 10^{-7}$ sec. Thus in order to change the field by 200 amp/m, from +100 amp/m to -100 amp/m, a time $\Delta t$ of the order of $3 \times 10^{-8}$ sec is required. Such changes of field can be accomplished by available pulse techniques. One should not forget that the complete reversal of the field, from $+H$ to $-H$, should be completed during a time that is much shorter than the spin-lattice relaxation time $\tau_1$.

Up to the present this method of inversion for obtaining negative spin temperatures was only used in conjunction with nuclear spin systems, but it played on important part in the history of introduction of the concept of negative temperatures. One can hope that this method will also become useful in construction of paramagnetic oscillators of very high frequency.

## 18. EXAMPLES OF PRACTICAL REALIZATION OF TWO-LEVEL SYSTEMS [14, 40, 41]

A characteristic property of two-level masers is their pulsed mode of operation and instability of the gain or magnitude of the generated power during the duty cycle. The duty cycle is preceded by a time interval $t_1$ during which the inversion takes place. The interval $t_1$ should be much smaller than the spin-lattice relaxation time $\tau_1$. After this there follows the duty cycle $t_2$, which is usually also a small fraction of the time $\tau_1$.

The longest time interval is the interval $t_3$ during which the system returns to thermal equilibrium with the lattice. The interval $t_3$ is "useless" because during this interval there is no amplification (or oscillation). In order to make the maser more efficient, it is desirable to somehow shorten the time $t_3$ required for reestablishment of equilibrium. Several methods for shortening the interval $t_3$ were proposed. For example, if impurity-doped semiconductors are used as the working substance the time $t_3$ is shortened by illumination of the semiconductor with visible light. One of the most popular substances used in two-level masers is silicon with impurities consisting of elements of group V (P, As, Sb) or lithium. The electron paramagnetic resonance in such semiconductors is caused by electrons that are localized on the donor atoms and interact with these atoms by Coulomb forces. At liquid helium temperatures and with a concentration of the order of $10^{17}$-$10^{18}$ donor atoms per cm$^3$ the relaxation time $\tau_1$ is of the order of one minute. If the sample is illuminated with visible light, free conduction electrons appear, causing the spin-lattice relaxation time to decrease to about 1 ms. Thus a flash of light following immediately after the interval $t_2$ renders the "dead time" negligibly small, and as a result the maser can operate almost continuously.

A second method for decreasing the "useless" time interval $t_3$ is based on the following considerations. Crystals containing two types of paramagnetic systems are chosen as the working substance. The spin-lattice relaxation times of these systems must be as different as possible. Systems with longer relaxation times are used for amplification and oscillation. After the period of amplification or oscillation is completed the static magnetic field is changed in such a way as to make the separation between the spin levels of the utilized system equal to the separation between any pair of spin levels of the second system. In this case an exchange of energy between the spin systems of the first and second type becomes possible. The second spin system will quickly transmit its energy to the lattice. The efficiency of this mechanism increases as the spin-lattice relaxation time for the second system decreases.

Also interesting is the proposition to use mechanically conveyed devices for continuous replenishment of the supply of excited paramagnetic centers. For example, a scheme was proposed in which the working substance is placed on a rotating disc. This disc successively enters two resonant cavities in one of which the inversion takes place and in the other the amplification or generation of microwaves.

The first attempt to construct a two-level maser was made by Combrisson, Honig and Townes in 1956. The working substance was silicon with phosphorus impurities. Three crystals were placed in a rectangular resonant cavity, designed for a frequency of 9000 Mc/s and having an unloaded $Q_0 = 10,000$. The volume of the resonant cavity was 4.4 cm$^3$ and the volume of the working substance 0.6 cm$^3$. The entire system was cooled to 2°K. A klystron served as the signal source and its power output was applied to one arm of a T-bridge. The resonant cavity was connected to one of the other arms of the bridge; i.e., a maser with a reflection resonant cavity was constructed. The inversion of levels was accomplished by adiabatic fast passage.

This attempt at maser construction, however, was not successful inasmuch as the oscillatory regime could not be obtained. Therefore the experimenters limited themselves to measurements of the power required to maintain the electro-magnetic field in the cavity and of the fraction of power radiated by the working substance during spin transitions. From these measurements the concentration of donor atoms was determined. For various samples it was shown to be about $0.3$–$0.6 \times 10^{17}$ donor atoms per $cm^3$.

In order to obtain amplification it is necessary that the average radiated power $P$ exceed the power losses $P_w$ in the cavity walls. The power $P_w$ is equal to

$$P_w = \frac{\omega_0 V \mu_0' H_c^2}{Q_0} \tag{72}$$

where $Q_0$ is the unloaded $Q$ of the resonant cavity; $V$ is the volume of the cavity; $H_c^2$ is the value of the square of the mag-netic field, averaged over the volume of the cavity; and $\mu_0'$ is the magnetic permeability of vacuum.

From the condition $P > P_w$ we obtain on the basis of Eq. (43)

$$N > \frac{kTV\mu_0' H_c^2}{\omega_0 Q_0 \mu^2 \tau_2 H_1^2 V_m} \tag{73}$$

where we let $|T_s| = T$. The experimental conditions of Com-brisson, Honig and Townes were such that $1/\tau_2 = \Delta \nu = 4$ Mc/s and $H_c^2 \cong H_1^2$. After substitution of these values the condition (73) becomes: $N > 10^{17}$ donor atoms per $cm^3$. This requirement was not satisfied and therefore the expected effect was not achieved.

It appears that the situation can be corrected by increasing $N$. But with an increase in $N$ the spin-spin interaction increases and the relaxation time $\tau_2$ decreases. It can be seen from (73) that a decrease in $\tau_2$ requires another increase in $N$. Conse-quently, this method is only slightly effective.

Other researchers (Feher, Gordon, Buehler, Gere and Thurmond [44]) chose another approach. Natural silicon consists of two isotopes, $Si^{28}$ and $Si^{29}$. The $Si^{29}$ nuclei are paramagnetic and interact with phosphorus atoms. This causes broadening of the EPR line. Therefore an isotopically pure (99.88 ± 0.08%) silicon with atomic weight of 28 was taken as the working substance; the width of the EPR line was decreased from 215 amp/m to 17.5 amp/m. At the temperature of 1.2 °K used by these authors, the spin-lattice relaxation time was 60 sec. The level inversion was obtained by adiabatic fast passage. The working substance was a crystal with a volume of 0.3 cm³. The donor concentration was $N = 4 \times 10^{16}$ cm⁻³, the transition frequency was 9000 Mc/s, and the unloaded $Q$ of the cavity at 1.2 °K reached 20,000. With these conditions the oscillatory regime was obtained. The maximum pulse amplitude obtained corresponded to a radiated power of 2.5 $\mu$w. After 100 $\mu$s the pulse amplitude decreased to 0.1 of its maximum value.

Amplification and oscillation were also obtained with other working substances. In particular, Chester, Wagner and Castle [45] used single crystals of quartz, as well as magnesium oxide, for construction of a reflection resonant-cavity maser. When these substances are irradiated by neutrons, $F$ centers appear and the electrons localized in these $F$ centers cause paramagnetic resonance. The authors used a resonant cavity whose unloaded $Q$ was of the order of 6000; the operating frequency was 9000 Mc/s and the temperature 4.2 °K. The level inversion was obtained by adiabatic fast passage. The length of the excitation pulses was 50-100 $\mu$s with a power level in the pulse of the order of 0.5 w.

When quartz was used, the number of active centers was of the order of $10^{18}$ and their radiation after adiabatic passage excitation persisted for 2 $\mu$s so that amplification was observed for 1.2 $\mu$s. The product $B\sqrt{G}$ was of the order of $5\times10^6$ Mc/s at amplification levels which varied from 21 to 8 db. The oscillatory regime was also obtained. The peak oscillatory pulse power reached 12 mw and the oscillations lasted about 10 $\mu$s.

In the experiments with magnesium oxide the number of paramagnetic centers was of the order of $10^{17}$ and the inverted state was preserved for about 2.5 $\mu$s. About 125 $\mu$s after inversion the gain was 20 db and after 720 $\mu$s it decreased to 3 db.

The above-mentioned disadvantages of two-level masers—pulsed mode of operation, synchronization of the operating time with the inversion period, and changes in gain and output power during the operating time—create serious difficulties in practical applications of such systems. On the other hand, two-level masers open up new possibilities for obtaining sources of comparatively high power in the millimeter and submillimeter wavelength regions. An increase in the static field "spreads" the levels apart so that the generated frequency and, correspondingly, the output power are increased. The solution of such problems requires large pulsed fields, resonant cavities with very high $Q$, and more suitable working substances.

Together with solid paramagnetic substances, paramagnetic gases can also be utilized in two-level masers. In gaseous substances, however, the relaxation processes are much less pronounced and the spectral lines are very narrow. Therefore, gas masers are useful as standards of frequency and time.

One of the first devices of this type was the atomichron (cesium frequency standard [46]). Cesium vapor is the working substance in this maser. When the beam of cesium atoms passes through an inhomogeneous field the atoms are sorted into two groups (the electron angular momentum is equal to $\frac{1}{2}$). The atoms in the upper energy level enter a resonant cavity which is also connected to a quartz oscillator. When the cavity frequency and the transition frequency are equal, induced radiation of the atoms takes place and they decay to the lower energy level. After emerging from the cavity the beam again passes through an inhomogeneous magnetic field and all atoms that could possibly have remained in the upper level are extracted from the beam. The atoms in the lower level are incident on a detector (a heated wire). The atoms incident on the detector are converted to ions. These ions impinge on an electron multiplier whose output current regulates a servosystem which can correct the frequency of the quartz oscillator which in turn excites the resonant cavity. An atomichron uses the hyperfine structure line of cesium (9192.631830 Mc/s) and the line width is 200 c/s. A frequency stability of $10^{-11}$ was achieved, which corresponds to 0.1 sec in 300 years. The nonuniformity of the rotation of earth was determined more accurately with the help of the atomichron. At the end of October the rotation slows down by 0.53 sec and at the end of May it accelerates by 0.065 sec. The dimensions of an atomichron are approximately $2.1 \times 0.5 \times 0.5$ m.

Even greater stability is obtained with an atomic hydrogen maser [47]. The atoms are also sorted with the help of an inhomogeneous magnetic field. The cavity ($Q = 60,000$) is tuned to the frequency of the hyperfine transition of 1420.405 Mc/s. The minimum intensity of the atoms in the beam required to maintain oscillation is $4 \times 10^{12}$ particles/sec.

## Chapter VI

## Three-Level Masers

### 19. METHODS OF EXCITATION [40, 41]

If a paramagnetic system has more than two energy levels $(S > \frac{1}{2})$, there appear new possibilities for application of such systems in amplification and generation of microwaves. This possibility was first pointed out by Basov and Prokhorov in 1955 [67] and was realized in 1957 by Scovil, Feher and Seidel [49].

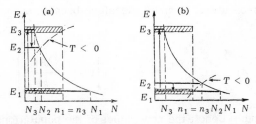

FIG. 44. Energy level system used in a simple method of pumping in a three-level quantum amplifier

Let us examine the physical principles on which the operation of three-level masers is based. Figure 44 shows two cases of possible distribution of energy levels $E_1, E_2$ and $E_3$. At

137

thermal equilibrium between the spin system and the lattice the populations $N_1$, $N_2$ and $N_3$ of the levels are given by the Boltzmann distribution, i.e.,

$$N_3/N_1 = e^{-\frac{(\Delta E)_{31}}{kT}} \quad \text{and} \quad N_2/N_1 = e^{-\frac{(\Delta E)_{21}}{kT}} \quad (74)$$

where

$$(\Delta E)_{31} = E_3 - E_1 \quad \text{and} \quad (\Delta E)_{21} = E_2 - E_1 \quad (75)$$

Assuming $(\Delta E)_{21} < (\Delta E)_{31} < kT$, if we expand the exponents into series and retain only the first-order terms we get

$$N_3 = N_1 \left[ 1 - \frac{(\Delta E)_{31}}{kT} \right] \quad (76)$$

$$N_2 = N_1 \left[ 1 - \frac{(\Delta E)_{21}}{kT} \right] \quad (77)$$

Let us now suppose that the paramagnetic system is subjected to an electromagnetic field whose frequency satisfies the condition $h\nu_{31} = (\Delta E)_{31}$ and whose intensity is such that the transition between the levels $E_1$ and $E_3$ is saturated. The populations of levels 1 and 3 therefore become equal. Thus we can write

$$n_1 = n_3 = \frac{N_1 + N_3}{2} \simeq N_1 \left[ 1 - \frac{1}{2} \frac{(\Delta E)_{31}}{kT} \right] \quad (78)$$

where $n_1$ and $n_3$ denote the new values of the populations in levels 1 and 3.

If at the same time the population $N_2$ of the middle level differs from $n_1 = n_3$, i.e., $N_2 > n_1$ or $N_2 < n_3$, conditions that allow continuous radiation will have been established. In order to be able to utilize the transitions between the third and the second level for amplification, $n_3$ should exceed the population $N_2$. Thus, according to (77) and (78), it is necessary to satisfy the inequality

$$\frac{1}{2} (\Delta E)_{31} < (\Delta E)_{21}$$

In other words, the middle level should be nearer to level 3 than to level 1; i.e., the frequency $\nu_{31}$ of the pumping field should exceed the frequency of the amplified signal by more than a factor of 2 (Fig. 44a). Similarly, for amplification at the transition frequency between levels 2 and 1 it is necessary to have $N_2 > n_1$. This gives

$$(\Delta E)_{21} < \frac{1}{2} (\Delta E)_{31}$$

It is seen that the results are the same: the frequency $\nu_{31}$ of the pumping field exceeds the frequency of the amplified signal by a factor of more than 2, and in this case level 2 should be located closer to level 1 than to level 3 (Fig. 44b).

The spin-lattice relaxation time between levels 1 and 3 should be sufficiently long, otherwise large power levels are required to saturate levels 1 and 3. A high pumping-radiation power level is undesirable for a number of reasons. In particular, at a high power level the maser will heat up and its efficiency will immediately be reduced.

The relaxation times $(\tau_1)_{32}$ and $(\tau_1)_{21}$ should also satisfy some definite conditions. For the case shown in Fig. 44a they should satisfy the relationship

$$(\tau_1)_{32} > (\tau_1)_{21}$$

If the relaxation time between levels 1 and 2 is large, the population of level 2 will start to increase and the difference in populations $(n_3 - n_2)$ will decrease. The reverse is also true: a short spin-lattice relaxation time $(\tau_1)_{21}$ assures the transition of "spent" atoms from level 2 to level 1, returning the system to its initial state, with the number of atoms distributed among the levels according to the Boltzmann distribution. An increase in population $N_1$ allows one to obtain (at saturation) a large population $n_3$ of the third level. It is obvious that an increase in $n_3$ and a decrease in $n_2$ will favorably affect the performance of the maser.

The relaxation time $(\tau_1)_{32}$ should be sufficiently long. If this is not so, the number of nonradiative transitions from level 3 to level 2 will be large. This will cause a decrease in population $n_3$ and an increase in the population of the second level. The population excess $(n_3 - n_2)$ will decrease and the performance of the maser will become less efficient.

For the case corresponding to Fig. 44b, analogous reasoning leads to the requirement that the relaxation time $(\tau_1)_{32}$ should be shorter than the relaxation time $(\tau_1)_{21}$. Satisfaction of this requirement will assure a sufficiently large population excess $(n_2 - n_1)$ to enable the maser to work.

The requirement that the frequency of the auxiliary radiation (pumping frequency) should exceed the signal frequency by more than a factor of 2 is a definite limitation. For amplification of high frequencies a pumping source of even higher frequency is required, and at the same time the pump should have sufficient power to saturate the third level. From this point of view, "optical pumping" is of great interest. The essence of this method is as follows. Assume that we have an atom whose ground level is triply degenerate (orbital angular momentum $L = 0$, total angular momentum $J = 1$), as shown in Fig. 45. The figure also shows one of the excited levels which is fivefold degenerate ($L = 1$, $J = 2$). We will also assume that the atom is irradiated with circularly polarized light whose frequency corresponds to the separation between the investigated groups

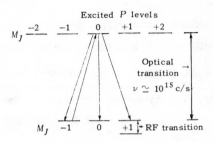

FIG. 45. System of energy levels used in "optical pumping"

of levels. Suppose that the polarization is such that each photon has a projection $M_J = +1$ (the projection of the angular momentum of a photon is referred to the same direction as the projection of the total angular momentum of the atom). When such a photon is absorbed, the atom changes its state from $J = 1$ and $M_J = -1$ to $J = 2$ and $M_J = 0$, according to the selection rule $\Delta M_J = +1$. The atoms in the upper levels, however, are strongly excited and radiate spontaneously, returning to their ground state (lifetime $\sim 10^{-8}$ sec). Polarization of photons with spontaneous emission is random (according to the rules $\Delta M_J = 0, \pm 1$). As a consequence, after emission the atom can assume any of the three states with $J = 1$. Many repetitions of similar cycles of transitions result in a decrease of the population of level $J = 1$ and $M_J = -1$ and an increase in the population of level $J = 1$ and $M_J = +1$. (The situation would be similar if we considered other possible transitions from remaining lower levels up and down. This is obvious from examination of Fig. 45 and consideration of the previously mentioned selection rules.) If now the lower level with $J = 1$ and $M_J = +1$, with its increased population, is connected by a microwave transition with a still lower level, the conditions required for radiation could be created.

The most suitable materials for optical pumping are the alkali metals and atomic hydrogen. Optical pumping uses the transition between the atomic $^2S^{1/2}$ ground state and the first excited $P$ state. The hyperfine structure of the ground state, caused by nuclear spins, is used for RF transition.

This method for obtaining radiating systems with the help of optical pumping has a number of limitations. First, the method is apparently only applicable to gaseous systems. Second, the pressure of the gas should be relatively low, otherwise the unpolarized photons which are radiated by the excited atoms will be reabsorbed by unexcited atoms before these photons leave the gaseous system. Since these photons are not polarized, all levels participate in their absorption and subsequent radiation, with the result that the population of the level used for the RF transition is drastically reduced. Finally, a large line width in the excited $P$ states also prevents the accumulation of a significant excess of atoms in the upper level of the RF transition. In particular, the transitions between the excited $P$ levels and the lower level of the RF transition become possible. All these factors necessitate a decrease in bandwidth in order to obtain amplification. For this reason, amplifiers with optical pumping are most applicable in situations in which a narrow bandwidth is not a disadvantage, e.g., in oscillators with high spectral purity and in time standards.

Finally, let us consider three more possible methods of supplying pumping radiation. These methods are based on the use of four levels [40, 48]. One of these, called the method of effective doubling of the frequency of pumping radiation (push-push method), can be used in situations in which the positions of the energy levels are such that the separations between the first and second level and between the second and the fourth level are identical (Fig. 46a). Supplying the pumping radiation at the frequency $\nu_{21} = \nu_{42}$ and obtaining saturation $(n_1 = n_2 = n_4)$ a condition can be reached in which the population of the fourth level $n_4$ exceeds the population of the third level $N_3$. In this case the transitions between the fourth and third levels can be used for amplification or oscillation.

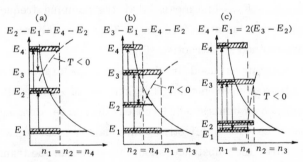

FIG. 46. System of four energy levels used for pumping; a—push-push method, b—push-pull method, c—method in which the pumping frequency is lower than the signal frequency

The second method, called the method of symmetric excitation (push-pull method), can be used in situations in which the separations between the first and third levels and between the second and fourth levels are identical (Fig. 46b). Supplying the pumping radiation at the frequency $\nu_{31} = \nu_{42}$ and obtaining saturation ($n_3 = n_1$ and $n_4 = n_2$), a state can be reached in which the population of the third level $n_3$ exceeds the population of the second level $n_2$. Consequently, the transition 3—2 can be used for amplification or generation of microwaves.

Even more remarkable is the case when the levels are spaced in such a way that $E_3 - E_2 = \frac{1}{2}(E_4 - E_1)$. With application of the pumping radiation at the frequency $\nu_{32}$ it is possible in a number of cases to obtain the saturation of transitions 2—3 and 1—4. It is furthermore possible that at the same time the population of the third level will exceed the population of the first level. This excess will be larger if the spin-lattice relaxation time $(\tau_1)_{34}$ is made shorter. If the probability of induced transitions 3—1 is sufficiently large, this transition can be used for amplification and generation of microwaves. Since $E_3 - E_1 > E_3 - E_2$, this means that the pumping frequency is lower than the radiation frequency. This is mainly what makes the above method so attractive (Fig. 46c).

The substances used in three-level masers must satisfy some definite requirements. The number of spin levels of the paramagnetic systems should be no less than three. Consequently, all substances with an effective spin equal to $\frac{1}{2}$ must be excluded. Furthermore, the substance should be such that the transitions at the pumping frequency are allowed. It may be seen from Figs. 44 and 46 that for such transitions the quantum number of the projection of the angular momentum on the direction of the static magnetic field should change by a quantity $\Delta M = \pm 2$. However, it was noted in the discussion of electron paramagnetic resonance that such transitions, as a rule, are forbidden.

Actually, if it is assumed that the different spins do not interact, the Hamiltonian of the spin system should contain only the Zeeman energy: each magnetic moment interacts only with the external magnetic field $H$ and each state with energy $E_m$ is described by a wave function $\psi_m$. Such a representation of the spin system, however, is only approximate. In reality there exist spin-spin interactions: "magnetostatic" dipole-dipole interaction and "dynamic" spin-exchange processes. In addition, the state of an atom with energy $E_m$ will also depend on the state of the adjacent atoms (and these can even be in some other angular momentum states). As a consequence the state of the investigated atom with energy $E_m$ should be described by the wave function

$$\psi = \psi_m + a_1 \psi_{m-1} + a_2 \psi_{m+1} + a_3 \psi_{m-2} + a_4 \psi_{m+2} + \cdots$$

Computation of the matrix elements that determine the transition probability yields the result that together with the transitions $\Delta M = \pm 1$ there appear transitions $\Delta M = \pm 2, \pm 3$, etc. It also becomes evident that the static and high-frequency magnetic fields do not necessarily have to be perpendicular to each other; orientation at other angles is also possible. The number and probability of such transitions in an individual case will be determined by the specific properties of the substance, the nature of its interaction with the magnetic field, the orientation of the static magnetic field with respect to the axis of symmetry of the crystal, etc.

A strong spin-spin interaction causes splitting of the energy levels even in the absence of an external magnetic field. The larger the amount of splitting in a zero external field, the larger the spin-spin interaction and the more probable the "forbidden" transitions. Therefore, when choosing the material for a three-level maser attempts are made to obtain an initial splitting of the order of $h\nu_{31}$. For small values of the initial splitting the probability of transitions at the pumping frequency $\nu_{31}$ is small and it is difficult to achieve saturation.

The forbidden transitions appear also in the presence of quadrupole interaction ($\Delta M = \pm 1, \pm 2$) and with multiple quantum transitions [14]. (Multiple quantum transitions involve simultaneous absorption of several quanta by a paramagnetic atom.)

## 20. POWER OUTPUT [14, 40]

Let us investigate the power output of a three-level maser. For clarity we will assume the distribution of energy levels shown in Fig. 44a. The induced radiated power $P$ is equal to the product of the difference in populations of the levels used for amplification, $n_3 - n_2$, the quantum of energy $h\nu_{32}$, and the induced-transition probability $W_{32}$:

$$P = (n_3 - n_2) h\nu_{32} W_{32} \tag{79}$$

Let us compute the quantity $n_3 - n_2$. We will assume that at steady state the number of transitions to each level from the other levels is equal to the number of transitions from the given level to the other levels. Formulating this condition for each level, we obtain a system of three equations:

$$n_2 w_{21} + n_3 w_{31} + n_3 W_{31} = n_1 w_{13} + n_1 W_{13} + n_1 w_{12}$$

$$n_1 w_{12} + n_3 w_{32} + n_3 W_{32} = n_2 w_{21} + n_2 w_{23} + n_2 W_{23}$$

$$n_1 w_{13} + n_1 W_{13} + n_2 w_{23} + n_2 W_{23} = n_3 w_{31} + n_3 W_{31} + n_3 w_{32} + n_3 W_{32}$$

where $n_1$, $n_2$ and $n_3$ are the populations of the levels (under operating conditions), $w_{ik}$ are the transition probabilities (per unit time) from level $i$ to level $k$ caused by thermal lattice vibrations, and $W_{ik}$ are the probabilities of induced transitions from level $i$ to level $k$ (where $W_{ik} = W_{ki}$). Let us examine the last equation in some detail. On the left side of this equation are the transitions that increase the population of the third level and on the right are those that decrease the population of the third level. The first and third terms on the left correspond to spontaneous transitions from the first and second levels to the third level, the second term is the number of induced transitions from the first level to the third caused by the presence of the pumping field, and the fourth term is the number of induced transitions from the second level to the third due to the presence of an amplified signal. The first and third terms on the right side represent the number of spontaneous transitions to the first and second levels, and the second and fourth terms on the right are the numbers of induced transitions to the same levels caused by the presence of the pumping field and of the signal.

The meaning of all the terms in the two remaining equations can be explained in a completely analogous matter (the first equation expresses the equilibrium condition for transitions of the first level, the second equation, that for the second level). Let us note that these three equations are not independent. Thus, subtracting the third equation from the first we obtain the second equation, subtracting the second from the third we obtain the first, and adding the first and second we obtain the third. This is because no matter what the distribution of spins among the various levels, the total number of spins remains constant. Therefore, in subsequent discussion we will only consider the second and third equations (although any other pair could also serve the purpose). These equations, together with the condition of conservation of the number of spins, make up the system

$$n_1 w_{12} - n_2 (W_{32} + w_{23} + w_{21}) + n_3 (W_{32} + w_{32}) = 0$$

$$n_1 (W_{13} + w_{13} + w_{12}) - n_2 w_{21} - n_3 (W_{13} + w_{31}) = 0$$

$$n_1 + n_2 + n_3 = N$$

Eliminating $n_3$ from the first two equations we obtain

$$n_1 (w_{12} - W_{32} - w_{32}) - n_2 (2W_{32} + w_{32} + w_{23} + w_{21}) +$$
$$N (W_{32} + w_{32}) = 0$$

$$n_1 (2W_{13} + w_{13} + w_{31} + w_{12}) - n_2 (w_{21} - W_{13} - w_{31}) -$$
$$N (W_{13} + w_{31}) = 0$$

Since the intensity of the saturating field with frequency $\nu_{31}$ is quite large, the probability of induced transitions $W_{13}$ is much larger than the probabilities of spontaneous transitions $w_{ik}$. With this condition the second equation becomes

$$2n_1 + n_2 = N$$

It follows from this that $n_1 = n_3$ (we have verified the known fact that at saturation the populations of the levels are the same). In addition, substituting $n_2 = N - 2n_1$ into the first equation gives the following result for $n_1$:

$$n_1 = \frac{N (W_{32} + w_{23} + w_{21})}{3W_{32} + 2w_{23} + w_{32} + 2w_{21} + w_{12}}$$

Correspondingly, the expression for $n_2$ becomes

$$n_2 = \frac{N (W_{32} + w_{32} + w_{12})}{3W_{32} + 2w_{23} + w_{32} + 2w_{21} + w_{12}}$$

Knowing $n_1$ and $n_2$ we can find the difference of level populations:

$$n_1 - n_2 = n_3 - n_2 = N \frac{w_{23} + w_{21} - w_{32} - w_{21}}{3W_{32} + 2w_{23} + w_{32} + 2w_{21} + w_{12}}$$

Since $w_{ik} \approx w_{ki}(1 - h\nu_{ki}/kT)$, where it is assumed that $h\nu_{ki}/kT \ll 1$ [Section 3, Eqs. (3) and (4)], we can let $w_{32} \simeq w_{23}$ and $w_{21} \simeq w_{12}$ in the denominator of the last expression. This cannot be done in the numerator because the numerator contains only the difference $w_{ik} - w_{ki}$ and hence the quantities $h\nu_{ki}/kT$ acquire an independent and determining character. With the above considerations, we obtain the difference in populations of the energy levels:

$$n_1 - n_2 = n_3 - n_2 = \frac{hN}{3kT} \frac{w_{21}\nu_{21} - w_{32}\nu_{32}}{W_{32} + w_{32} + w_{21}} \tag{80}$$

Remembering now Eq. (79) for the induced radiated power and using (80) we get

$$P = \frac{hN}{3kT} \frac{w_{21}\nu_{21} - w_{32}\nu_{32}}{W_{32} + w_{32} + w_{21}} h\nu_{32}W_{32} \tag{81}$$

With large input signals, when the probability $W_{32}$ is much larger than the sum of the probabilities $w_{32}$ and $w_{21}$, Eq. (81) can be simplified to

$$P = \frac{h^2 N\nu_{32}}{3kT} (w_{21}\nu_{21} - w_{32}\nu_{32}) \tag{82}$$

It should be remembered, however, that an excessive increase of the input signal can cause a decrease in the difference $n_3 - n_2$ and thus a decrease in the output power (due to saturation).

Analysis of the expression for the power shows that there are several ways in which the output power can be increased.

1. *Increase of the difference* $w_{21}\nu_{21} - w_{32}\nu_{32}$. With comparable relaxation times this condition corresponds to a large value of $\nu_{21} - \nu_{32}$; i.e., the frequency $\nu_{21}$ should be much larger than the frequency $\nu_{32}$. This means that the frequency of the pumping radiation $\nu_{31}$ should be much higher than that of the amplified signal $\nu_{32}$.

The difference $w_{21}\nu_{21} - w_{32}\nu_{32} = \dfrac{\nu_{21}}{(\tau_1)_{21}} - \dfrac{\nu_{32}}{(\tau_1)_{32}}$ will also increase when the relaxation time $(\tau_1)_{32}$ becomes larger with respect to the relaxation time $(\tau_1)_{21}$. A large value of the relaxation time $(\tau_1)_{32}$ is also very desirable because then the main cause of transitions from level 3 to level 2 will be the signal field. In other words, the inversion of the energy levels will be used mainly to perform useful work. A small value of the relaxation time $(\tau_1)_{21}$ is also desirable because then the lifetime of the atoms in the energy state $E_2$ will be small and the atoms will decay relatively quickly to level 1. An increase in the population of the first level at saturation (when $n_1 = n_3$) will cause an increase in the difference $n_3 - n_2$. It should be remembered, however, that a paramagnetic crystal is a macroscopic system which possesses a number of mutually related properties. A change in some of these properties can induce changes in others. In particular it appears that a decrease of the relaxation time $(\tau_1)_{21}$ usually causes a decrease in the relaxation time $(\tau_1)_{31}$. Consequently, use of crystals with small $(\tau_1)_{21}$ necessitates an increase in the pumping power required for saturation. If this power becomes large the crystal heats up internally and the efficiency of the maser decreases.

Therefore it is necessary to use some optimum relaxation time $(\tau_1)_{21}$.

2. *Increase of the total number of available spins* N. Physically this condition is obvious. There is, however, a limit to the increase of N. When the volume of the sample increases, the dielectric losses also increase. An increase in the number of spins in a given volume will cause an increase in spin-spin interaction. At the same time the relaxation time $\tau_2$ decreases and the molecular bandwidth increases. On the other hand, it is known that the molecular bandwidth significantly influences the overall bandwidth of the maser. It is also known that an increase in maser bandwidth causes a decrease in amplification. It appears then that in order to obtain large gain it is desirable to increase the number of spins N and simultaneously decrease the relaxation time $\tau_2$. Since these requirements are mutually contradictory, an optimum combination should be selected.

3. *Decrease in temperature* T. This condition is based on the fact that at low temperature the population of the lower levels is higher (in the nonoperating mode). Accordingly, the inversion results in a large population difference $n_3 - n_2$. Furthermore, a decrease in temperature decreases the noise level of the maser.

4. *Increase in operating frequency* $\nu_{32}$. This condition, however, requires a corresponding increase in the frequency of the pumping field $\nu_{31}$. In this sense the use of noncoherent sources of pumping radiation even in the optical region of the electromagnetic spectrum is especially desirable.

In conclusion we will make one further remark regarding the power of the pumping radiation. This power should be sufficiently large to overcome the spin-lattice relaxation and complete the saturation. However, it should not be so large as to, first, cause overheating of the paramagnetic substance and, second, prevent the establishment of an equilibrium inside the spin system. The second requirement is related to the fact that at a large amplitude of the alternating magnetic field the probability of induced transitions $W_{13}$ can become larger than the probability of spin-spin exchange ($\sim \tau_2^{-1}$). Then a coherent ordering of the spin system will take place and, due to spin correlation, the probability of transitions will increase drastically (the system will go into a "superradiant" state), which will cause saturation of the entire line. Therefore, in order to avoid nonlinear effects, a pumping power is required such that the reorientation time of the spins is larger than the spin-spin relaxation time. (The pumping power should, however, be such that the reorientation time of the spins is not greater than the spin-lattice relaxation time. Otherwise saturation will not be achieved.)

The value of the absorbed pumping power was computed by Bloembergen; its order of magnitude is [49]

$$P_a \cong \frac{N h^2 \nu_{31}^2 w_{13} \gamma H_1 \tau_2}{3 kT} \tag{83}$$

where $H_1$ is the strength of the RF magnetic field whose frequency is $\nu_{31}$.

## 21. THREE-LEVEL RESONANT-CAVITY MASERS

The description of quantum amplifiers and oscillators of the resonant-cavity type given in Section 16 applies equally to three-level resonant-cavity masers. The only change that must be considered is a different way of computing the magnetic figure of merit $Q_m$. In Section 16 the magnetic figure of merit is computed starting with the average radiated power obtained from the excited two-level system. In three-level systems $Q_m$ must be computed from the expression for radiated power derived in the preceding section. In all other cases, when the functional variation of $Q_m$ is not considered, the analysis given in Section 16 is also applicable to three-level resonant-cavity masers. However, this in no way diminishes the differences that exist between two-level and three-level masers. Two-level solid-state masers can work only in a pulsed mode; three-level masers, as a rule, work in a continuous mode. For a two-level resonant-cavity maser the cavity is designed to resonate at only one frequency; for a three-level resonant-cavity maser the resonant-cavity maser the cavity is designed to resonate at and signal frequency). In the case of three-level masers new requirements are imposed on the sample material. The orientation of the static magnetic field with respect to the high-frequency field and the crystal axes becomes a very important parameter. Without discussing all these questions at length, we will make only a few remarks concerning the noise figure of resonant-cavity masers [40, 41].

The noise power at the output of a resonant-cavity maser is determined by the noise power delivered to the maser through the input waveguide, the thermal radiation of the resonant cavity walls, and the spontaneous radiation of the paramagnetic crystal. The first two components, which determine the noise of a maser, can be expressed as the thermal radiation density [see Eq. (55)]. The spontaneously radiated power (per unit bandwidth) can also be formally represented in an analogous way:

$$P_{ps} = \frac{h\nu}{e^{\frac{h\nu}{kT_s}} - 1} = \frac{N_2}{N_1 - N_2} h\nu \qquad (84)$$

where $T_s$ is the spin temperature, and $N_1$ and $N_2$ are the populations of the levels considered.

Using the transmission line and the resonant-cavity theory, one can show that in case of large gain ($G \gg 1$) the noise figure of a resonant-cavity maser with a transmission cavity is given by the expression

$$F \simeq \left[ 1 + Q_1 \left( \frac{1}{Q_0} + \frac{1}{Q_2} \right) \right] \left[ 1 - \frac{P_{ps}}{P_p} \right] \qquad (85)$$

(notation as in Section 16). For a resonant-cavity maser with a reflection cavity the noise figure has the form

$$F \simeq \left[ 1 + \frac{Q_1}{Q_0} \right] \left[ 1 - \frac{P_{ps}}{P_p} \right] \qquad (86)$$

In both cases the first factor accounts for the losses in the system and the second factor is related to the noncoherent spontaneous radiation. Both formulas are very similar and if in the expression for the noise figure of the transmission-cavity maser we let $Q_2^{-1} = 0$ we obtain the formula for the noise figure of the reflection-cavity maser. Moreover, comparison of the two expressions shows that for identical loaded $Q_L$ the noise figure of a transmission-cavity maser will always be higher than that of a reflection-cavity maser. This is because in a transmission cavity the induced emission power is not fully utilized since a portion of this power goes to the input waveguide.

Formulas (85) and (86) can be somewhat simplified if we note that the figure of merit $Q_0$ of a resonant cavity is much greater than the coupling figures of merit $Q_1$ and $Q_2$. Furthermore, in a transmission-cavity maser the input coupling is usually made stronger (low $Q_1$) than the output coupling (high $Q_2$). Under these conditions, and considering that $h\nu \ll kT$ and $h\nu \ll k|T_s|$, both formulas take the same form, which is identical to Eq. (56).

In a two-level maser the absolute value of the spin temperature $T_s$ is the same as the temperature of the resonant cavity or, in extreme cases, exceeds this temperature by a negligible amount (under conditions of full inversion and low power of amplified signal). With three-level systems the situation changes somewhat because in this case the inversion of the levels does not mean their complete reversal.

Using the notation and the results of the preceding section we compute the spin temperature $T_s$ for the three-level case. The temperature $T_s$ is introduced by the relationship

$$\frac{n_3}{n_2} = e^{-\frac{h\nu_{32}}{kT_s}} \simeq 1 - \frac{h\nu_{32}}{kT_s} \tag{87}$$

Substituting into this the values of $n_3$ and $n_2$ obtained in Section 20 and neglecting the transition probability $W_{32}$ we obtain

$$T_s = -\frac{T\nu_{32}(w_{32} + w_{21})}{(w_{21}\nu_{21} - w_{32}\nu_{32})} \tag{88}$$

The ratio $|T_s|/T$, which can be obtained from this expression, is called the inversion coefficient and is sometimes used in the analysis of the performance of three-level masers.

Thus the noise figure of a three-level resonant-cavity maser is

$$F \approx 1 + \frac{T}{T_0} \frac{(w_{32} + w_{21})\nu_{32}}{(w_{21}\nu_{21} - w_{32}\nu_{32})}$$

where $T_0$ is room temperature.

Examination of this formula shows that the noise will be smaller when the factor $w_{21}\nu_{21} - w_{32}\nu_{32}$ is made larger. Possible methods for increasing this difference were discussed in detail in the preceding section. The noise also decreases with a decrease in the operating temperature $T$ and with a decrease in the signal frequency.

Let us make a numerical estimate of the spin temperature $T_s$ and the noise figure $F$, assuming reasonable values for the various parameters that define these two quantities. Suppose that $T = 3°K$, $T_0 = 300°K$, $\nu_{32} = 3000$ Mc/s, $\nu_{31} = 10,000$ Mc/s (i.e., $\nu_{21} = 7000$ Mc/s) and $(\tau_1)_{21} \approx (\tau_1)_{32}$. The computation gives $T_s = -4.5°K$ and $F = 1.015$ (slightly over 0.06 db). The low internal noise of quantum oscillators and amplifiers is their basic outstanding feature.

## 22. THREE-LEVEL TRAVELING-WAVE MASERS [40, 49]

The general analysis of waveguide type masers, given in conjunction with the two-level maser, and the results obtained there are largely applicable to three-level waveguide masers. The basic difference is that in deriving specific expressions for gain, noise figure, etc., it will be necessary to consider all three levels.

Let us investigate the gain coefficient and the bandwidth of a traveling-wave maser (TWM). We will begin with expression

(50). In our case the specific form of the quantity $C$, which depends on the spin temperature, relaxation times, etc., will of course be different than in the case of a two-level maser.

The bandwidth is defined as the frequency range between the half-power points. If the gain is expressed in db then the bandwidth is defined by the points at which the gain is 3 db smaller than the center frequency gain $G(0)$ (these definitions are equivalent).

Expressing $G(\nu)$ in db at the center frequency as well as at the limits of the bandwidth, dividing the first by the second, and assuming $G$ sufficiently large we obtain

$$B\sqrt{G} = 3^{1/2}\Delta\nu \tag{89}$$

The gain coefficient can generally be expressed in terms of the magnetic figure of merit $Q_m$. For a traveling-wave system

$$Q_m = \omega \frac{W dx}{dP} \tag{90}$$

where $W$ is the RF energy stored per unit length of the structure, and $dP$ is the power loss in the length $dx$.

For a maser in an operating mode the power should increase, since the "losses" should be negative. Therefore in subsequent discussion we will use only the absolute value of $Q_m$. The total power flow is equal to the product of the energy stored per unit length and the group velocity $v_g$, i.e.,

$$P = W v_g$$

Using (90) and introducing the slowing factor $s = \dfrac{c}{v_g}$, we obtain after integration

$$P = P_{\text{in}} \exp\left[\frac{\omega S l}{c\,|Q_m|}\right] \tag{91}$$

Expressing the length of the structure by the number $N$ of free space wavelengths which can be fitted into this length, we write the expression for the power gain in the form

$$G = \exp\left[\frac{2\pi SN}{|Q_m|}\right] \tag{92}$$

Expressing the gain in decibels, we obtain

$$G = 27.3 \frac{SN}{|Q_m|} \tag{93}$$

At the same time the bandwidth $B$ becomes

$$B = \Delta\nu\left(\frac{|Q_m|}{9.1SN}\right)^{\frac{1}{2}} \tag{94}$$

A waveguide maser has a number of advantages compared with a resonant-cavity maser. These are as follows.

1. In a resonant-cavity maser the bandwidth is determined by the resonant properties of the cavity and is usually much narrower than the width of the EPR line. In a waveguide maser the resonant systems are absent. A slow-wave maser system can have a bandwidth that is much larger than the resonant line of the utilized crystal; i.e., the bandwidth $B$ of a traveling-wave maser will be essentially limited only by the properties of the paramagnetic crystal. It follows from this that TWM's have a broader bandwidth than do resonant-cavity masers.

2. The broad-band feature of a slow-wave TWM system allows considerable electronic tuning of the operating frequency by changing the pumping frequency and the value of the constant magnetic field. In resonant-cavity masers, with their sharp tuning to the center frequency, such retuning is impossible.

3. The output power of masers depends on the volume of the amplifying substance. In resonant-cavity masers this volume is an order of magnitude smaller than in TWM's. Consequently, waveguide masers have higher gain and power output.

4. Waveguide masers are less sensitive to fluctuations in the populations of the levels of the amplifying substance. These fluctuations cause changes in the magnetic figure of merit and lead to instability in gain. One cause of changes in level population is fluctuation of the pumping power. Due to the feedback present in resonant-cavity masers, fluctuations in pumping power will lead to regenerative effects and to a significantly larger gain instability than in a TWM, in which the positive feedback is much weaker.

5. Resonant-cavity masers are very sensitive to changes in the load. These variations cause changes in the feedback and lead to gain instabilities. Thermal expansion and vibrations which affect the coupling elements also cause instability. In a TWM all these factors have much weaker effects.

6. Waveguide masers are easily constructed as bandpass devices with separate input and output. Unidirectionality of quantum amplifiers is achieved by simple inclusion into the slow-wave system of elements with nonreciprocal attenuation. Hence there are no isolators or circulators to serve as additional noise sources.

## 23. PRACTICAL THREE-LEVEL MASERS

1. A description of the first continuously working three-level maser was first published in 1957 [49]. The sample was diluted gadolinium ethylsulfate. [In diamagnetic lanthanum ethylsulfate hydrate $La(C_2H_5SO_4)_3 \cdot 9H_2O$ crystals, 0.5% of the

lanthanum ions were replaced by gadolinium ions.] The spin of
the $Gd^{3+}$ ion is equal to 7/2 so that there are eight Zeeman
levels in the magnetic field $H$ (Fig. 47). If the alignment of the
RF magnetic fields is 45° and a static magnetic field of 235
ka/m is exactly perpendicular to the crystal axis, then the
separations between levels 2—4 and 2—3 correspond to frequencies of 17,520 Mc/s and 9000 Mc/s. The frequency $\nu_{42}$ was

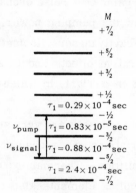

FIG. 47. Energy levels of $Gd^{3+}$
ion in a diluted crystal of gadolinium ethylsulfate ($H$ = 235 ka/m,
$\nu_{pump}$=17,520 Mc/s, $\nu_{signal}$ = 9000
Mc/s)

used for pumping and $\nu_{32}$ for amplification. In order to obtain
a large population in level 3, it is desirable to have $(\tau_1)_{43} \ll (\tau_1)_{32}$.
A decrease in time $(\tau_1)_{43}$ was achieved by addition of 0.2% of
cerium. The separation between the spin levels of $Ce^{3+}$ under
these conditions appeared to be equal to the interval $E_4 - E_3$
for the $Gd^{3+}$ ion. Due to the resonant spin-spin interaction of
gadolinium and cerium, the spin levels of $Ce^{3+}$ are excited by
the $4 \to 3$ transitions in gadolinium. Since the spin-lattice relaxation time of $Ce^{3+}$ is several orders of magnitude shorter
than that of $Gd^{3+}$ (for $Gd^{3+}$ at liquid helium temperatures

$\tau_1 \sim 10^{-4}$ sec), the excited cerium ions quickly give up their energy to the lattice. A decrease in the relaxation time $(\tau_1)_{43}$ permits lowering the frequency of the pumping radiation. In this example $\nu_{pump}$ exceeds $\nu_{signal}$ by less than a factor of 2.

The crystal described above (weight 90 mg) was placed in a two-frequency resonant cavity whose configuration is shown in Fig. 48a. The figure of merit of the cavity at a frequency of 9000 Mc/s was about 1000. This low $Q$ was the result of a spurious mode of oscillations that was excited at a frequency of 17,520 Mc/s.

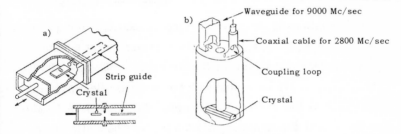

FIG. 48. Two types of doubly resonant cavity

The pumping power required to obtain amplification had to be at least 60 mw. An oscillatory mode was obtained with pumping powers of 100 mw or higher. The output power (at a frequency of 9000 Mc/s) was of the order of 20 $\mu$w.

Gadolinium ethylsulfate has a number of disadvantages as a maser material. The $Gd^{3+}$ ion has eight levels, of which only three are useful, and gadolinium ethylsulfate is chemically unstable.

2. In 1958 there appeared a description of a three-level maser which used diluted potassium chromicyanide as the sample [40]. [In potassium cobalticyanide $K_3Co(CN)_6$, 0.5% of the cobalt ions were replaced by chromium ions. This

corresponds to a spin concentration of $3.9 \times 10^{19}$ spins/cm$^3$.]
The spin of the $Cr^{3+}$ ion is equal to 3/2; its energy levels are
shown in Fig. 49.

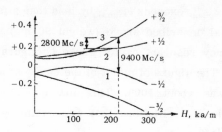

FIG. 49. Energy levels of a $Cr^{3+}$ ion in a
diluted crystal of potassium chromicyanide [50]

An exceptionally long spin-lattice relaxation time ($\tau_1 \sim 0.2$
sec at $1.25\,°K$) allows one to obtain saturation at a comparatively low pumping power of the order of 1-30 mw at the frequency $\nu_{31}$. At 175 ka/m and with the magnetic field parallel to
the $c$ axis of the crystal the pumping frequency is 9400 Mc/s
and the signal frequency $\nu_{32}$ is 2800 Mc/s. The static magnetic
field was applied perpendicularly to the axis of the coaxial
resonator (Fig. 48b). With the use of loop coupling the fundamental coaxial line mode was excited in the resonant cavity at a
frequency of 2800 Mc/s and a higher $TE_{113}$ mode at 9400 Mc/s.
The choice of the modes is determined by the attempt to obtain
the lowest cross coupling between the feeders at 2800 and 9400
Mc/s. The figure of merit of the resonant cavity at the signal
frequency is about 23,000. The pumping power was supplied
through a waveguide and was coupled through an iris. The input
and output signals were separated with the use of a 30-db
directional coupler. The crystal (approximately 1/10 the volume
of the resonant cavity) was placed at one end of the resonant
cavity and the cavity was enclosed in a double Dewar.

The gain of the maser was determined from the attenuation required to maintain a constant signal amplitude at the input of a spectrum analyzer. When the pumping power was increased from 1 to 30 mw the gain increased from 12 to 31 db and the bandwidth changed from 430 to 50 kc/s. The product $B\sqrt{G}$ remained approximately constant at about 1.8 Mc/s. A stable gain of 37 db was obtained with a bandwidth of only 25 kc/s. Since the natural width of the absorption line of the paramagnetic crystal is of the order of 30-50 Mc/s, it is obvious that the bandwidth is completely determined by the loaded $Q$ of the resonant cavity.

With suitable changes in the coupling $Q$ of the resonant cavity an oscillatory mode was achieved at a frequency of 2800 Mc/s. For a saturating power of the order of 0.35 mw the output power was roughly 0.05 $\mu$w. The maximum efficiency of the power transformation $P_{signal}/P_{pump}$ is achieved with $P_{pump} = 1$ mw and is only 0.14% (-28.5 db), for which the output power is 1.4 $\mu$w. When the pumping power is increased above 6 mw the output power increase becomes very small, reaching a maximum of 4 $\mu$w at $P_{pump} \approx 10$ mw (Fig. 50). The measured noise temperature of the maser was of the order of 20°K.

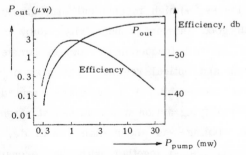

FIG. 50. Power output and efficiency as a function of pumping power [50]

It is interesting to compare these experimental results with theoretical computations. The initial data for the computation are as follows: $N = 1.9 \times 10^{19}$ spins/cm$^3$, $\tau_2 = 0.6 \times 10^{-8}$ sec, $\nu_{pump} = 9400$ Mc/s, $\nu_{signal} = 2800$ Mc/s, $T = 1.25\,°$K, and $V_{crystal} \sim 0.1\,V_{cavity}$. Assuming that the probability of induced transitions at the signal frequency is much larger than the probability of thermal transitions and letting $\tau_1 \sim 0.2$ sec everywhere we obtain a radiated power of the order of 7 $\mu$w, magnetic $Q_m \sim -2000$, $B\sqrt{G} = 2\nu_{signal}|Q_m^{-1}| \sim 2.5$ Mc/s (for computational formulas see Sections 17 and 20). The experimental and theoretical results are in good agreement.

3. The development of quantum paramagnetic amplifiers and oscillators has been characterized by an intensive search for better materials and attempts to increase the bandwidth, decrease the dimensions of the maser, increase the operating temperature, etc. The usefulness of corundum $Al_2O_3$ with a chromium impurity (ruby) as a maser material proved advantageous. Ruby is characterized by the following favorable properties: 1) convenient magnitude of initial splitting, corresponding to frequencies most often used in practice, 2) adequately long spin-lattice relaxation time ($\sim 0.1$ sec at $4.2\,°$K), 3) low dielectric losses, 4) high mechanical strength and chemical stability, 5) good thermal conductivity. Ruby masers were constructed for all frequencies, from centimeter waves to the infrared and optical regions. The size of a number of masers was decreased. A ruby maser was described in which the amplifying action was obtained at a temperature of $195\,°$K; the cooling was accomplished with dry ice [49]. A small magnet, placed together with the resonant cavity in a cryostat, was the source of the constant magnetic field.

The overall dimensions of the maser were of the order of 15 × 15 × 30 cm.

Finally, let us mention one more resonant-cavity maser which works in the wavelength region of 5-6 mm [51]. It uses rutile ($TiO_2$) with an $Fe^{3+}$ impurity (0.12% by weight) as the sample material. The energy levels are shown in Fig. 51. The excitation is by the push-pull scheme. The pumping frequency

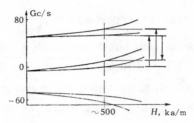

FIG. 51.  Energy levels of an iron ion $Fe^{3+}$ in rutile [51]

is about 78.2 Gc/s and the signal frequency varied from 49 to 57 Gc/s when the magnetic field was varied from 440 to 570 ka/m. A characteristic property of this maser is that variation of the magnetic field within the above limits does not change the relative spacing of the utilized levels so that the pumping frequency remains constant. A large initial zero-field splitting (43.3 and 81.3 Gc/s) allows the use of moderate magnetic fields. The resonant cavity was a cube with an edge of 1.5 cm, and the crystal volume was 0.15 $cm^3$. The operating temperature was 1.6°K. Maser action was observed with 50-mw pumping power.

4. From the appearance of the first masers in 1957 their technical characteristics have been continually improved. While in the first masers the product of the square root of the gain and the bandwidth was ∼ 0.5 Mc/s, at the present there are masers in which this product is several hundreds of megacycles.

This progress is closely connected with the appearance and development of waveguide quantum amplifiers.

The gain of quantum masers is directly proportional to the slowing factor. Therefore the problem of development of slow-wave structures for waveguide masers is of great importance.

All slow-wave structures can be roughly divided into three groups: dielectric, geometric and resonant. The slowing factor of dielectric systems is proportional to $\sqrt{\frac{\epsilon}{\epsilon_0}}$. In order to obtain a slowing factor of 100, substances with a relative dielectric constant of the order of 10,000 are required. Such substances, as a rule, are sensitive to temperature changes, unstable and have large losses. As a consequence the dielectric slowing method for waveguide masers is not very promising.

A helix is an example of a geometric slowing system. The slowing factor of a spiral is $D/L$, where $D$ is the diameter of the helix and $L$ is the length of the line from which the helix is made. If the diameter of the helix is 1 cm and the length is 10 cm, a pitch of the order of 0.3 mm is required to obtain a slowing factor of about 100. With such a small pitch, coupling exists between the windings and the system exhibits a sharp resonance. Furthermore, a helix is not suitable for other reasons: 1) the field decreases rapidly with the distance from the helix and the coupling between the field and the substance rapidly decreases; 2) in an ordinary helix the direction of polarization of the RF field does not preserve its orientation but rotates along the helix. Therefore an ordinary helix is not suitable for application in a TWM.

The most widely used slow-wave structures in waveguide masers today are those of resonant type. Such structures have a resonant frequency so that the maser must operate within their bandwidth. By proper choice of resonant elements the bandwidth can be made sufficiently large to exceed the width of the EPR line. As an example, Fig. 52 is a schematic representation of a comb-type slow-wave structure and its magnetic field distribution.

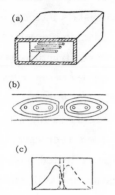

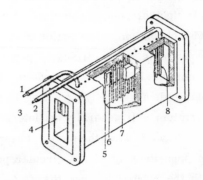

FIG. 52. Comb-type slow-wave structure and its magnetic field distribution: a—waveguide with slow-wave structure, b—distribution of magnetic field, c—amplitudes of the circularly polarized components [49]

FIG. 53. Traveling-wave maser: 1—input signal, 2—output signal, 3—coaxial cables, 4—pump waveguide, 5—ruby with 1% chromium serving as an isolator, 6—aluminum oxide spacer, 7—ruby with 0.05% chromium serving as amplifying substance, 8—probe coupling to cable

The comb structure is well suited to use in TWM's. Its field distribution is such that on one side of the rods the propagated wave is right-circularly polarized and on the other side it is left-circularly polarized. This property can be used to obtain unidirectional damping (so as to amplify the wave propagated in one direction only). Let us examine this, using one of the first waveguide masers as an example [49]. This maser is shown schematically in Fig. 53. A ruby containing 0.05% chromium was used as the working substance. It was found that a pumping power of ~ 100 mw, which is sufficient to obtain saturation in a ruby with 0.05% chromium, is not sufficient to saturate the pumping transition if the chromium concentration is of the order of 1%. If both of these types of ruby are placed in the slow-wave structure then the ruby with the lower concentration will produce amplification while the ruby with the higher concentration will produce attenuation. Both crystals interact only with right-circularly polarized radiation. Placing the ruby with 0.05% chromium in the region of the slow-wave structure where the wave is right-circularly polarized and the ruby with 1% chromium in the region where the wave is left-circularly polarized, an amplifying action is achieved. The ruby with 1% chromium will generally not interact with the direct wave. For the reflected wave the picture changes. The reflected wave has the opposite sign of the circular polarization and will interact with the ruby which has the greater chromium concentration. It is obvious that absorption will take place. By this method a unidirectional attenuation system is obtained: in the forward direction there is only amplification and in the reverse direction only attenuation.

In the traveling-wave amplifier described above (Fig. 53) the forward gain was 23 db and the reverse attenuation was 29 db (at a temperature of 1.5°K). When the constant magnetic field was shut off the losses in the structure were 3 db. The bandwidth of the slow-wave system was 350 Mc/s (from 5750 Mc/s to 6100 Mc/s). The maser bandwidth at the frequency of 5800 Mc/s was 25 Mc/s. In order to scan the entire bandwidth of the slow-wave structure by electronic tuning the pumping frequency had to be varied from 18,900 Mc/s to 19,500 Mc/s and the constant field from 313 to 324 ka/m. The measured noise figure of the above maser is equal to 1.037 (or 0.16 db).

5. The most important feature of quantum amplifiers is the exceptionally low level of internal noise. This influences the applications of the amplifiers: radio frequency spectroscopy, radio astronomy, radar and communications. In radio frequency spectroscopy masers are desirable in the search for and investigation of weak lines. In radio astronomy they can be of great value in reception and studies of radio frequency noise from galactic and extragalactic sources. Measurements of atmospheric noise were carried out with a traveling-wave maser [52]. It appeared that when the antenna is directed at the zenith the atmospheric noise temperature is $2.5 \pm 0.75\,°K$ (calculation gave $2.7\,°K$).

The research on application of masers in long-range radar is especially intensive. Utilization of masers allows a significant increase in sensitivity and maximum range of radar. From this point of view the following example is typical. The use of a maser as a preamplifier in a radar system has allowed lowering the total noise temperature (of maser, circulator, mixer and IF amplifier) to $65\,°K$ [43]. The noise temperature of the entire receiver (including waveguides, TR and anti-TR, rotating joints, etc.) was $173\,°K$. Without a maser the noise temperature of a radar receiver is of the order of $1500-2500\,°K$. The application of quantum amplifiers in communications is also promising, especially the utilization of masers in space communications.

The main ideas and considerations of the maser concept formed a basis for the development of new devices—quantum oscillators and light amplifiers (also called "lasers"). Lasers work on the same principle utilized in three-level masers. The working substance is often a pink ruby. A diagram of the optical levels of ruby is shown in Fig. 54. With intense irradiation of ruby with green light ($\sim$ 5000 A), the $Cr^{3+}$ ions become

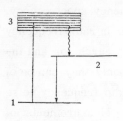

FIG. 54. Energy levels used in a laser

excited. The lifetime in the excited state is very short and the ions drop to level 2 by means of a nonradiative transition. The transition $2 \to 1$ is from a metastable stage; the atoms drop from this level to the lower one, radiating red light (6943 A). The end faces of the ruby crystal are polished to make them precisely parallel. The ends are coated with silver, one end only partially (transmission coefficient of several percent). A characteristic property of lasers is a very small radiation angle (for a ruby it is of the order of $2 \times 10^{-2}$ radians) [53].

Present-day lasers operate in the region from 7000 A to the infrared (2.5 $\mu$). The lower end of the spectral range can apparently be extended. There are no good reasons that should prevent the construction of lasers even in the millimeter wavelength region. The upper bound of the spectral range is determined by the level of spontaneous radiation. (In masers the main noise source is thermal radiation, but this is absent in lasers.)

To complete our discussion of quantum paramagnetic oscillators and amplifiers, we will remark that the advent of masers has in some sense created a new stimulus in science and technology. Solid-state research was accelerated, cryogenic techniques and RF instrumentation were improved, and new low-noise devices have appeared (e.g., parametric amplifiers). The potentialities of masers have only begun to be realized.

## Chapter VII

# Radio Frequency Spectroscopy of Gases

## 24. ROTATIONAL LEVELS OF LINEAR MOLECULES
### [13, 54, 55]

In order to explain the mechanism of absorption of radio frequencies and microwaves by gas molecules it is necessary to investigate the energy levels of a molecule and the transitions that are possible between these levels. In the simplest case a molecule consists of two atoms. Each level of the gross electronic structure of such a molecule is a function of the principal quantum numbers of the separate, isolated atoms as well as of the distance between the atoms, i.e., $E_{el} \propto U(r)$.

Furthermore, $E_{el}$ will be dependent on the vibrational motion of the atoms. It is not difficult to find the energy $E_{vib}$ of the vibrational motion; almost every text on quantum mechanics has a solution for the energy levels of an oscillator. Thus $E_{vib}$ has the form

$$E_{vib} = h\nu \left( v + \frac{1}{2} \right) \tag{95}$$

where $\nu$ is the proper frequency of molecular vibrations and $v$ is the vibrational quantum number, which assumes a discrete number of values $(0, 1, 2, \ldots)$.

In addition to the vibrational motion, a diatomic molecule can also have another type of motion—rotational. The rotational energy of a rigid rotator is equal to

$$E_{\text{rot}} = \frac{1}{2} mr^2 \dot{\phi}^2$$

where $I = mr^2$ is the moment of inertia of the molecule and $\dot{\phi}$ is the angular speed of rotation. Introducing into this expression the angular momentum $P = mr^2 \dot{\phi}$ we obtain

$$E_{\text{rot}} = \frac{1}{2} P \frac{P}{mr^2} = \frac{P^2}{2mr^2}$$

Since $P^2 = \left(\frac{h}{2\pi}\right)^2 J(J + 1)$, where $J$ is the rotational quantum number, we have

$$E_{\text{rot}} = \frac{h^2 J(J + 1)}{8\pi^2 I} \tag{96}$$

Therefore the energy of a diatomic molecule including vibration and rotation is

$$E_{n,v,J} = U(r) + \frac{h\omega_{\text{vib}}}{2\pi} \left(v + \frac{1}{2}\right) + \frac{h^2 J(J + 1)}{8\pi^2 I} \tag{97}$$

where $n$ is a quantum number characterizing the term.

The rotational-vibrational structure of energy levels is shown schematically in Fig. 1.

The rotational energy levels are often written in the form

$$E_{\text{rot}} = hBJ(J + 1) \tag{98}$$

where $B = h/8\pi^2 I$.

This relationship not only is valid for diatomic molecules but also describes equally well the rotational levels of linear polyatomic molecules. The transition frequencies between adjacent rotational levels are given by the relationship

$$\nu = \frac{h}{4\pi^2 I}(J + 1) \tag{99}$$

From this relationship we see that the frequencies in the rotational spectrum are determined by a single molecular parameter—the moment of inertia. The larger the moment of inertia, the smaller will be the factor in front of the parentheses $(J + 1)$ and the larger the number of rotational transitions falling in the microwave region. For example, the cyanogen bromide molecule CNBr has a moment of inertia $I = 204 \times 10^{-40}$ gram·cm$^2$. The rotational transition frequencies equal $\nu = 0.8226 \times 10^{10} \times (J + 1)$ c/s, from which $\nu_1 = 8226$ Mc/s, $\nu_2 = 16,452$ Mc/s, $\nu_3 = 24,678$ Mc/s, etc. (a total of 37 lines in the wavelength range down to 1 mm). The hydrocyanic acid molecule HCN has a moment of inertia $I = 18.9 \times 10^{-40}$ gram·cm$^2$. Consequently, the frequencies of the lines are $\nu_1 = 88,800$ Mc/s, $\nu_2 = 147,600$ Mc/s, $\nu_3 = 266,400$ Mc/s, etc. Starting with $J = 3$ the lines will be in the wavelength region below 1 mm. If a molecule is still lighter and its moment of inertia is smaller than $15 \times 10^{-40}$ gram·cm$^2$, no lines of the rotational spectrum will be in the microwave region.

The rotational levels predicted by Eq. (98) agree well with experimental data. In a more precise calculation, however, it is necessary to obtain a more accurate solution of the problem of the molecular energy levels. For this purpose it is usually sufficient to solve the problem in second-order perturbation theory. This solution is as follows:

$$E_{\text{rot}} = h[BJ(J+1) - DJ^2(J+1)^2] \tag{100}$$

where $B = h/8\pi^2 I$, $D = 4B^3/\omega^2$ for a diatomic molecule. For a triatomic molecule $D = 4B^3 \, (\xi_1^2/\omega_1^2 + \xi_2^2/\omega_2^2)$, where $\omega_1$ and $\omega_2$ are the proper frequencies of the longitudinal vibration, and $\xi_1$ and $\xi_2$ are constants. The correction term $DJ^2(J+1)^2$ becomes important only at large values of $J$, i.e., when the rotational energy of the molecule is large and there are large centrifugal forces tending to distort the molecule. Therefore it is often said that the correction term in Eq. (100) accounts for centrifugal distortions.

Our discussion is not yet complete. First, the electronic structure levels discussed above are spatially degenerate. The vibrational and rotational structures do not give rise to fine structure levels. Actually the energy levels depend on the rotational quantum number $J$, which defines the total angular momentum $P = \dfrac{h}{2\pi} \sqrt{J(J+1)}$. The orientation of the angular momentum $\vec{P}$ in space, however, is entirely arbitrary; i.e., the energy levels do not depend on the quantum number $m$, which takes on $2J+1$ values. Therefore the energy levels have a spatial degeneracy of multiplicity $2J+1$.

Second, the expression for the energy levels obtained above agrees well with experiment only for singlet terms, when the total spin $\vec{S}$ of the molecule is zero. Actually, if $S = 0$, no further corrections to the energy levels are necessary because the spin–orbit interaction is absent. When the total spin differs from zero, the spin–orbit interaction must be taken into account.

In addition to spin–orbit interaction there is spin–spin interaction. However, its effect is considerably smaller. The influence of other types of interaction, such as spin-rotation, orbit-rotation, etc., is even less.

In conclusion let us note that in linear polyatomic molecules it is necessary to take into account the so-called bending vibrations. In a diatomic molecule only one longitudinal vibration is possible. Consider a triatomic molecule. It is evident that the number of possible longitudinal vibrations is two. Moreover, in addition to the longitudinal vibrations there are also two possible transverse vibrations: in the plane of the drawing (Fig. 55) and perpendicular to the drawing. Thus for a triatomic molecule the number of possible normal modes of vibration is four. (A normal mode of vibration is a motion in which all the coordinates of the displacement of the atoms in a molecule vary sinusoidally at the same frequency.) A linear molecule consisting of $n$ atoms has $3n - 5$ normal modes.

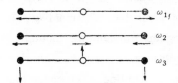

FIG. 55. Normal modes of a linear triatomic molecule

The total vibrational energy of the system in the case of a linear triatomic molecule consists of the energy of all possible normal modes of vibration, i.e.,

$$E_v = h\nu_1\left(v_1 + \frac{1}{2}\right) + h\nu_2\left(v_2 + \frac{1}{2}\right) + h\nu_3\left(v_3 + \frac{1}{2}\right) \qquad (101)$$

The description of vibrational levels of a triatomic molecule in this form, however, is somewhat crude. Levels determined by expression (101) are degenerate. In fact, the transverse vibrations $\nu_3$ in two mutually perpendicular planes can be regarded as independent, and a linear vibration in any plane will be the result of the superposition of two simple vibrations. At the same time it is necessary that the phases of these vibrations be identical. If the phases of two "initially" simple harmonic motions are different then after superposition of such motions we will find out that the atoms will describe neither ellipses nor circles. Thus, in general, every atom of a linear triatomic molecule has, in addition to the longitudinal vibrations, a rotational motion (around the intermolecular axis) resulting from the vibrations and giving rise to a constant angular momentum $\vec{l}$ along this axis. The rotational energy levels of a triatomic molecule are independent of the direction of this vibrational angular momentum and are therefore doubly degenerate (since the vector $\vec{l}$ can have two equivalent directions). Actually the degeneracy is approximate. It exists only because we neglected the interaction between the rotational angular momentum and the vibrational angular momentum. Inclusion of this interaction results in symmetrical splitting of rotational levels. This splitting is called $l$ doubling. The magnitude of $l$ doubling for $\Pi$ vibrational states ($l = \pm 1$) can be expressed by the formula

$$\Delta\nu = q J (J + 1) \qquad (102)$$

where

$$q = (v + 1) \frac{B^2}{\omega} (1 + f)$$

($v$ is the quantum number of normal bending vibrations, $\omega$ is the frequency of degenerate normal bending vibrations, and $f$ is some function of the Coriolis forces).

Transitions between $l$-doublet levels were first discovered in the hydrocyanic acid molecule. The transition frequencies between the rotational levels of this molecule are given by the expression $\nu = 8.8800 \times 10^{10}(J + 1)$. For the transition from $J = 6$ to $J = 7$ we have $\nu_6 = 621,600$ Mc/s. The transition frequency between the $l$-type doublet levels for the rotational level $J = 6$ is 9460 Mc/s. This property, which makes it possible to observe transitions between the levels of $l$-type doubling in a given rotational level, also allows observations to be made in the microwave frequency region of internal transitions in levels with large $J$ near the top of the Boltzmann distribution. However, it also permits observation of absorption spectra of gases at low radio frequencies. For example, the transition with $\Delta J = 0$ between the levels of $l$-type doubling with $v_2 = 1$ and $J = 1$ for OCS gives an absorption line at 12.68 Mc/s.

## 25. ROTATIONAL LEVELS OF SYMMETRIC-TOP MOLECULES [13, 55]

In a symmetric-top molecule the moments of inertia with respect to the two principal axes are equal. While in a linear molecule the total angular-momentum vector $\vec{P}$ for $\Sigma$ states is always perpendicular to the axis of the molecule, in a symmetric-top molecule, even in the case of $\Sigma$ states, the vector $\vec{P}$ does not have to be perpendicular to the figure axis. In general, the vector $\vec{P}$ has a constant component along the figure axis (Fig. 56). This component is due to the motion of the heavy nuclei. (In linear molecules there can also be an angular-momentum component along the figure axis, but this component will be due to the motion of electrons. In a $\Sigma$ state, when $L = 0$, linear molecules have no angular momentum component along the axis.)

Rotational motion of symmetric-top molecules is complex. It consists of rotational motion of the molecule around a symmetry axis, which in turn precesses around the direction of the total angular momentum. The superposition of these two motions does not result in a simple rotation around the vector $\vec{P}$ (as might be suspected from the meaning of $\vec{P}$). The result is that vector $\vec{P}$ is not constant with respect to the molecule and that rotation takes place around some instantaneous axis whose position continuously changes. The situation can best be pictured by imagining a stationary cone whose axis is the vector $\vec{P}$ and a second cone, whose axis is the axis of the molecule, with the second cone rolling without slipping on the first. The motion of this second cone will correspond to the actual motion of the molecule. The line of tangency of the cones will be the instantaneous axis of rotation. This axis, as well as the axis of the molecule, rotates around the vector $\vec{P}$.

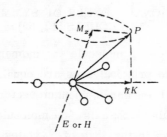

FIG. 56. Vector diagram for symmetric-top molecules

The rotational energy levels for a symmetric-top molecule are given by a formula that is very similar to the formula for rotational levels of linear molecules. In fact, according to classical mechanics, the kinetic energy of rotation of a rigid body is given by

$$E = \tfrac{1}{2} I_x \omega_x^2 + \tfrac{1}{2} I_y \omega_y^2 + \tfrac{1}{2} I_z \omega_z^2 = \frac{\vec{P}_x^2}{2I_x} + \frac{\vec{P}_y^2}{2I_y} + \frac{\vec{P}_z^2}{2I_z}$$

where $x$, $y$ and $z$ are the directions of the principal axes, and $I_i$, $\omega_i$ and $P_i$ are respectively the principal moments of inertia, angular velocity components and angular momentum components with respect to the axis $i$ ($i = x, y, z$). In our case $\vec{P}_z = \vec{K}$ and $\vec{P}_x + \vec{P}_y + \vec{K} = \vec{P}$. Denoting $\vec{P}_x + \vec{P}_y = \vec{N}$ (vector $\vec{N}$ is perpendicular to the vector $\vec{K}$), we have $\vec{P}_x^2 + \vec{P}_y^2 = \vec{N}^2$ and $\vec{N} + \vec{K} = \vec{P}$, so that $\vec{N}^2 + \vec{K}^2 = \vec{P}^2$. Therefore $\vec{P}_z^2 = K^2$ and $\vec{P}_x^2 + \vec{P}_y^2 = P^2 - K^2$. Denoting $I_z = I_A$ and $I_x = I_y = I_B$ we obtain

$$E = \frac{P^2}{2I_B} + \frac{K^2}{2I_B} + \frac{K^2}{2I_A}$$

In order to obtain the quantum-mechanical expression from the classical expression, we must subsitute: $\vec{P}^2 = \left(\frac{h}{2\pi}\right)^2 J(J+1)$ and $\vec{K}^2 = \left(\frac{h}{2\pi}\right)^2 K^2$. Therefore,

$$E = \frac{h^2}{8\pi^2 I_B} J(J+1) + \left(\frac{h^2}{8\pi I_A} - \frac{h^2}{8\pi^2 I_B}\right)K^2$$

or

$$E = h[BJ(J+1) + (A-B)K^2] \tag{103}$$

where $A = h/8\pi^2 I_A$ and $B = h/8\pi^2 I_B$. The second term in this expression is not constant and can take on different values corresponding to the different values of the quantum number $K$. However, since $P_z \equiv K$ is a component of the total angular momentum, the quantum number $K$ cannot be greater than the quantum number $J$, i.e., $J = K, K+1, K+2, \ldots$ According to Eq. (103) the states that differ only in the sign of $K$ have identical energy; i.e., all states with $K > 0$ are doubly degenerate. (Different signs of $K$ correspond to two opposite directions of rotation of the top.) Figure 57 shows the rotational energy levels for a prolate symmetric top [in a prolate top $I_A < I_B$; i.e., $A > B$ and the second term in (103) is positive].

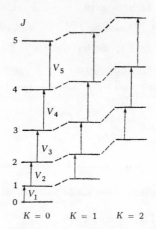

FIG. 57. Energy level diagram for a prolate symmetric top (all levels with $K > 0$ are doubly degenerate)

If we take into account the fact that the molecule is not rigid, it is necessary to introduce a correction for centrifugal distortion. This correction, which increases with increase in rotational energy, can be computed with the help of perturbation theory. It appears, however, that corrections computed by this method do not agree well with experiment. Therefore, instead of theoretical values of constants, the semiempirical coefficients $D_J$, $D_K$ and $D_{JK}$ are usually used. The first of these is connected with the precessional motion of the molecule as a whole and the second with the rotation of the molecule around the symmetry axis. The third is a coefficient arising from the cross term. With the correction for centrifugal distortion, the energy levels are given by the expression

$$\frac{E}{h} = BJ(J + 1) + (A - B)K^2 - D_J J^2 (J + 1)^2 - \qquad (104)$$

$$- D_{JK} J (J + 1) K^2 - D_K K^4$$

The transition frequencies of the absorption lines between adjacent rotational levels $J \to (J + 1)$, i.e., when $\Delta J = \pm 1$ and $\Delta K = 0$, will therefore be equal to

$$\nu = 2B(J + 1) - 4D_J (J + 1)^3 - 2D_{JK} (J + 1) K^2 \qquad (105)$$

Since the last two terms are considerably smaller than the first, the spectrum will consist of a series of lines separated by almost equal intervals $2B = \dfrac{h}{4\pi^2 I_B}$ (the spectrum will have the same form as in the case of linear molecules).

The main effect of the centrifugal distortion is that every line corresponding to a transition $J \to (J + 1)$ is split into $J + 1$ components (i.e., as many components as there are values of $|K|$). The separation between the adjacent components is usually of the order of 0.5 Mc/s.

In symmetric-top molecules, as in linear molecules, level splitting due to $l$-type doubling also occurs. The $l$-type doubling of levels can be observed as a splitting of lines in transitions between the rotational levels, as well as directly in absorption in transitions between the levels of $l$-type doublets $(\Delta J = 0)$.

As an example, Fig. 58 shows a diagram of the splitting of the rotational spectral line $J(2 \to 3)$ due to centrifugal distortion and $l$-type doubling.

Frequently gas molecules have a permanent electric dipole moment so that the interaction of the molecules with an RF field is, as a rule, electric; i.e., the electric dipole moments interact with the electric component of the RF field. It is also known that electric dipole transitions take place only when accompanied by a change in parity. Returning to Fig. 58, we note that two sublevels of $l$-type doubling with the same quantum number $J$ correspond to different wave functions: the lower level corresponds to a symmetric wave function and the upper to an antisymmetric one. Considering now the selection rules for $J$ in the case of dipole transitions ($\Delta J = 0, \pm 1$), as well as the selection rules for parity, we obtain the result that the line $2 \to 3$ is split into two components. Furthermore, direct transitions between the levels of $l$-type doubling of the same rotational

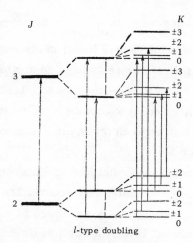

FIG. 58. Splitting of spectral line $J$ ($2 \to 3$) due to centrifugal distortion and $l$-type doubling

level are also possible (dotted lines in Fig. 58). Since the magnitude of the $l$-doublet splitting varies quadratically with $J$ [$\Delta\nu = qJ(J + 1)$], these transitions can be observed in the low radio frequency region (e.g., in OCS for $J = 1$, $v_2 = 1$, the transition frequency is 12.68 Mc/s), as well as in the high-frequency region for levels with large $J$ (e.g., in HCN for $J = 6$, $v_2 = 1$, the transition frequency is 9400 Mc/s).

The consideration of centrifugal distortion leads to still more splitting of the energy levels. The number of sublevels is determined by the possible values of the projection of the total angular momentum on the figure axis of the molecule. This number is equal to $K = 0, \pm1, \pm2, \ldots, \pm J$. Therefore each component of $l$-type doubling decomposes for $J = 2$ into three sublevels and for $J = 3$ into four sublevels (Fig. 58), so that the $2 \to 3$ spectral line will now have six components. In general the number of lines is equal to $2(J + 1)$, where $J$ is the quantum number of the lower level (it is assumed that $\Delta K = 0$).

## 26. INVERSION SPLITTING

For the energy levels of nonplanar molecules, including molecules of the symmetric-top type, there is another characteristic form of splitting which is absent in linear or planar molecules. This splitting is connected with inversion or "turning the molecule inside out." In fact, when inversion of a nonplanar molecule takes place, i.e., a reflection through the origin (center of gravity of the molecule), a new structure of the molecule is obtained which cannot be obtained by any other transformation of the molecule, such as rotation or translation.

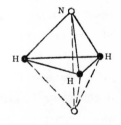

FIG. 59. The two structures of the ammonia molecule

As an example, we will investigate the ammonia molecule (Fig. 59). The nitrogen atom can take on two positions: above or below the plane containing the three hydrogen atoms. The transformation of one structure into the other is possible only as a result of passage of the nitrogen atom through the plane of the hydrogen

atoms. The potential energy curve for the nitrogen atom will
have two minima corresponding to the two stable configurations
of the molecule (Fig. 60). These minima are separated by a
potential barrier whose height is determined by the interaction

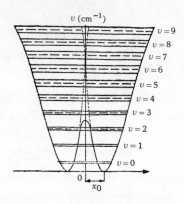

FIG. 60. Potential curve and some
energy levels of the $NH_3$ molecule

of the nitrogen atom with the hydrogen atoms during its passage
through the plane of the hydrogen atoms. Both configurations
have identical rotational levels because the moments of inertia
of both configurations obviously are the same. If we recall the
expression for the rotational levels of a symmetric-top molecule, we see that each level of the $NH_3$ molecule is doubly
degenerate. In general, this degeneracy, which is called inversion degeneracy, is a property of all nonplanar molecules.

In fact, the inversion degeneracy is only approximate and
can be removed if the finite height of the potential barrier is
considered. This can be adequately illustrated by considering
a particle in a square potential well with two minima, separated
by a rectangular barrier $U_0$ (Fig. 61). For simplicity we assume
that the side walls are infinitely high. The probability of the

particle being in regions 1 and 5 is zero ($\psi_1 = \psi_5 = 0$). The solution of the Schrödinger equation for regions 2, 3 and 4 has the form [56]

$$\psi_2 = A \sin kx$$

$$\psi_3 = B_1 e^{\kappa x} + B_2 e^{-\kappa x}$$

$$\psi_4 = C \sin k(2a + b - x)$$

where $k = 2\pi\sqrt{2mE}/h$ and $\kappa = 2\pi\sqrt{2m(U_0 - E)}/h$. With the requirement of continuity of the wave function and its derivative at $\kappa = a$ and $\kappa = a + b$, it can be shown that for a particle that lies sufficiently deep in the potential well ($\kappa \gg k$ and $\kappa b \gg 1$) the energy levels are given in first-order approximation by [56]

$$E_n = E_n^{(0)} - 2\frac{E_n^{(0)}}{\kappa a} \mp 4\frac{E_n^{(0)}}{\kappa a} e^{-\kappa b} \tag{106}$$

where $E_n^{(0)} = \dfrac{h^2}{8ma^2} n^2$, $n = 1, 2, 3, \ldots$

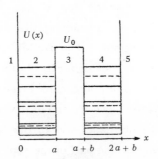

FIG. 61. Energy levels of a particle in a square-well potential with two minima

The quantity $E_n^{(0)}$ defines the energy levels of a particle in an infinite square well (with one minimum). The first two terms in (106) are independent of the barrier width $b$ and give approximate values for the energy levels of a particle in the left or right potential well (regions 2 and 4 in Fig. 61). In this approximation the energy levels (dotted lines) are doubly degenerate, corresponding to the possibility of finding the particle in region 2 as well as in region 4. Where we take into account the finite value of $b$, i.e., the possibility of the particle penetrating through the potential barrier, we obtain the splitting of levels. This splitting has an exponential dependence and increases with a decrease in the width of the barrier.

When the coefficients $B_1$, $B_2$ and $C$ are determined, it becomes clear that the upper sublevels correspond to antisymmetric wave functions $\psi_n(x)$ and the lower levels to symmetric wave functions.

In the case of real molecules, such as $NH_3$, the shape of the potential curve is, of course, not rectangular. The features of the level spacings and the characteristics of the inversion splitting, however, remain the same. The strength of the interaction between the levels in the "right" and the "left" potential wells, i.e., the size of the doublet splitting and, consequently, the probability of transition of a particle from one potential well to the other, are very small for levels located considerably below the top of the barrier and quite large for levels located near the top. For levels above the top of the barrier the doublet splitting continues to increase, and in the limit (far from the top) it becomes equal to one half of the separation between the adjacent unperturbed levels, which are obtained when the barrier is assumed to be impenetrable. The separations between these levels are twice as small as the initial separations and their eigenfunctions are alternately symmetric and antisymmetric.

The size of the inversion splitting is dependent to a large extent on the vibrational state of the molecule. The probability of a transition of the N atom through the plane of the H atoms in the $NH_3$ molecule is maximum when the vibrations of the N atom are perpendicular to the H plane (such motion is nearly obtained in the completely symmetric normal mode $\omega_2$). It is obvious that the inversion splitting will drastically increase with an increase in the vibrational quantum number. Thus for $\omega_2$ vibrations for $v = 0$ we have $\Delta\nu = 0.8$ cm$^{-1}$, for $v = 1$ we have $\Delta\nu \approx 36$ cm$^{-1}$, and for $v = 2$ we have $\Delta\nu \approx 312$ cm$^{-1}$. In the case of other normal modes the splitting is considerably smaller. For example, for $\omega_1$ vibrations we have $\Delta\nu \approx 1$ cm$^{-1}$ for $v = 1$ [57]. The inversion splitting depends not only on $v$ but also on the quantum numbers $J$ and $K$, since the potential energy of a molecule depends on the rotational state; the centrifugal forces change the distances between the atoms and, consequently, also the parameters of the potential curve (height and width of the barrier). When a molecule has a permanent dipole moment (for $NH_3$ this moment is equal to 1.46 debye), transitions between the sublevels of the inversion doublet are possible, and therefore a pure inversion spectrum of the $NH_3$ molecule should exist $(\Delta J = 0)$. At room temperature the majority of molecules are in the lowest vibrational level $(v = 0)$ and as a result only the frequencies of the inversion spectrum that correspond to this level are found in the radio frequency region. The above-mentioned splitting $\Delta\nu \approx 0.8$ cm$^{-1}$ corresponds to a wavelength $\lambda = 1.25$ cm.

## 27. GASEOUS RADIO FREQUENCY SPECTROMETERS
[12, 13, 55, 58]

The observation of the absorption spectrum of gases is not dependent on the presence of external fields. Therefore the basic elements of a gaseous radio frequency spectrometer are a microwave generator, an absorption cell and a detector. Figure 62 is a block diagram of a simple gaseous radio frequency spectrometer in which the measurements are performed as follows. The absorption by the absorption cell is measured at a given frequency with and without the gas and recorded with a suitable measuring device. After this the frequency of the

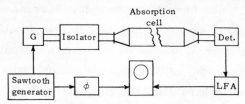

FIG. 62. Gaseous radio frequency spectrometer with frequency sweeping and a video receiver with a crystal detector

generator is retuned to some other value and the measurement is repeated, etc. The absorption curve is thus obtained from the measured points. In this method the cells can be used in waveguides or resonant cavities; the high density of the investigated gas often requires a magnetron as the signal generator. To increase the accuracy of the measurements it is necessary on the one hand to carefully stabilize the signal generator frequency and on the other hand to narrow the bandwidth of the amplifier. This is usually achieved by amplitude modulation of the signal frequency and narrow-band, synchronous detection. It is also necessary to make standing-wave ratio measurements in order to decrease possible errors. If a waveguide cell, shorted at one end, is used as the absorption cell, the attenuation caused by the gas can be obtained by measuring the difference in the standing-wave ratios with and without the gas.

It is obvious that fixed-frequency gaseous radio frequency spectrometers cannot be used to search for narrow lines. For this purpose the most simple and reasonably sensitive method is to sweep the generator frequency and observe the line on an oscilloscope screen (Fig. 62). Klystrons are generally used as the microwave signal source. They have a number of advantages compared with magnetrons: small dimensions, ease in handling, possibility of frequency tuning within relatively large ranges, reasonably high power output, and lower internal noise level. Frequencies in the millimeter and submillimeter regions are usually obtained with so-called harmonic generators, i.e., by multiplying the frequencies generated by the klystron to the required value with crystal converters. Of course, the power level of such a signal generator is usually quite small, but in a majority of cases it is sufficient for observations of the absorption spectrum. The problem of saturation is more severe in radio frequency spectroscopy of gases than in observation of magnetic-resonance spectra. This is because the sample being investigated is in the gaseous state (i.e., of low density) and the power incident on a molecule is thus several orders of magnitude higher than in magnetic-resonance radio frequency spectroscopy.

The low pressure of the investigated gas, mostly 1000-10 $n/m^2$ ( $\sim$10-0.1 mm Hg) and even lower, requires large absorption cells in order to obtain sufficient absorption for observation. The absorption cell is usually a section of waveguide with a cross section larger than that of the spectrometer waveguide; it is several meters long. The cell is connected with the main waveguide by tapered sections, integral with the cell, which serve to match the total impedance of the cell to the total impedance of the smaller, transmission waveguide. To simplify handling, the waveguide cell is sometimes made in the form of a spiral.

Sometimes a resonant cavity is used as the absorption cell. This gives easily observed absorption with relatively small dimensions. For efficient performance, however, the cavity must be tuned to the signal frequency, thus making it impossible to vary the frequency over wide limits. This gas spectrometer system, therefore, cannot be used for searching.

The low level of microwave power and even lower amplitude of the absorption signal in the radio frequency spectroscopy of gases leave the total internal noise level the main factor determining the sensitivity of any radio frequency spectrometer. The size of this noise level is usually determined by the internal noise of the detector and the amplifier. Therefore one of the main criterions for the choice of a detector or its mode of operation is its contribution to the total noise level.

Let us return now to the simple radio frequency spectrometer shown in Fig. 62. The frequency generated by the klystron is varied as a function of time by applying a low-frequency sawtooth voltage to the reflector of the klystron G. Such electronic tuning can be achieved in a range of 0.1% of the generated frequency while maintaining a linear relationship between frequency and time, and without power modulation. Thus the detector receives a modulated signal which after detection can be impressed on the vertical deflection plates of an oscilloscope whose sweep is synchronized with the frequency of the sawtooth generator. The absorption line will be displayed on the oscilloscope screen. By mechanical klystron tuning it is possible to change the generated frequency by about 15%. Therefore, by changing the frequency in steps, it is possible with the use of the above scheme to cover the entire frequency range of the spectrometer.

Faithful reproduction of the absorption signal requires as large a bandwidth as possible, with the bandwidth increasing as the pulse narrows. On the other hand, a wide bandwidth is completely undesirable. The reason is that, even disregarding the high noise level of the crystal detector at low modulation frequencies, when the frequency is swept large pulses of power arise in the waveguide line and appear at the detector. Such pulses originate as a result of reflections even with the most careful fabrication and matching of waveguide parts. This is because the matching conditions are dependent on frequency. If reflections appear near the ends of the absorption cell the maxima and minima will repeat every time the generator frequency changes by $c/2L$, where $L$ is the length of the cell. The longer the cell, the narrower are these power bursts. These bursts can very seriously obstruct the search for lines (especially weak ones). The basic measure adopted to decrease the effect of these bursts reduces to the following. A simple video receiver is connected directly to the crystal detector (it is sometimes called a crystal-video detector). At its output a filter is connected which drastically reduces the low frequencies so that the bandwidth of the amplifier is considerably decreased. The pressure of the investigated gas is selected in such a way that the width of the spectral line is considerably narrower than the undesirable bursts occurring due to reflections. Luckily, the magnitude of this pressure is not critical; for the majority of gases a pressure of 1 n/m$^2$ ($\sim 0.01$ mm Hg) is entirely satisfactory. At this pressure the width of the absorption line is usually about 1/10 of the width of the most narrow false burst peaks.

We stated above that for a more faithful reproduction of the absorption line for narrow pulses the bandwidth of the amplifier should be much wider than for wide pulses. Furthermore, it is known that the lower Fourier harmonics essentially determine the form of a pulse and that the higher harmonics decrease very quickly, the decrease being faster for wider pulses. Therefore in decomposition of a wide pulse one can limit oneself to a smaller number of harmonics. Now it becomes obvious that if the lower frequencies in the Fourier expansion are cut off for both pulses the result will be that the wider pulse will not go through the receiver, while the narrow pulse will be only insignificantly distorted.

The amount of lower harmonics to be cut off is determined as follows (for a given fixed bandwidth). If the modulation frequency is varied, the period of both signals (desired and false) will change but their relative width will remain the same. The Fourier spectrum will consequently increase or decrease and, at the same time, the amount of the lower frequencies (which are mainly responsible for the appearance of the wide pulses at the output of the spectrometer) eliminated by the filter will change. Therefore in gaseous radio frequency spectrometers of the type discussed here the frequency modulation of the klystron is optimized with respect to the picture observed on the oscilloscope screen.

One more advantage of this spectrometer is the ability to easily distinguish the absorption lines of parasitic signals. The absorption line signal will disappear immediately when the pressure of the gas in the absorption cell is increased 8-10 times, i.e., an amount required to sufficiently broaden the absorption line so that its main Fourier components fall in the region below the cutoff frequency of the filter.

The discussed radio frequency spectrometer is simple, rapid to use and reasonably sensitive. Some distortions introduced by the spectrometer, however, render it inapplicable to studies of line shapes.

In order to prevent an increase in sensitivity of a radio frequency spectrometer from being accompanied by a corresponding distortion of the spectral line shape, attempts are usually made in sensitive spectrometers to modulate the signal at a frequency of several tens or hundreds of kilocycles so that the signal can be amplified at these frequencies. A modulation frequency of the order of 100 kc/s is most often used.

There exist two methods for modulating the absorption signals: molecular modulation and source modulation. Both methods are similar in principle because in both the absorption line itself is used to obtain a high-frequency amplitude modulation of the microwave radiation. The simplest of such radio frequency spectrometers use a common narrow-band radio receiver in place of the amplifier.

The high-frequency source modulation method is as follows. A sweep voltage of much higher frequency (several Mc/s) from a modulator (a radio frequency generator) is superimposed on the slowly varying sawtooth voltage used to sweep the klystron frequency, causing the klystron frequency to make fast excursions up and down in frequency. At the same time the line is slowly swept across by the sawtooth sweep. Thus the absorption line acts as a discriminator (like the detector for frequency-modulated signals), converting the frequency modulation into amplitude modulation at the same frequency. Therefore, in order to amplify the signal after detection one can use a common radio receiver designed to receive amplitude-modulated signals.

For the normal operation of a radio frequency spectrometer the period of the high-frequency sweep voltage must be smaller than the traversal time through the line. In the opposite case the line shape will be considerably distorted and the sensitivity of the device substantially decreased. In this method special attention should also be paid to the depth of modulation. In order to observe as little distortion as possible in the line shape, the depth of modulation should be larger than the line width. For maximum sensitivity, however, the modulation depth should be much smaller than the line width.

If the period of the modulating voltage is smaller than the time required to traverse the line, it can be assumed that during the sweep time in a given portion of the line, the amplitude of the absorption signal in this portion will be modulated several times by the frequency of the modulation voltage. If furthermore the amplitude of the modulating voltage is negligible it can be assumed that the output signal will be sinusoidal and that its amplitude will be proportional to the slope of the absorption line at the given point of the sweep. Therefore, in a sweep period the oscilloscope screen will display a curve describing the first derivative of the absorption line. (It was established that for this purpose the amplitude of the modulation should not be greater than 1/6 of the distance between the points of maximum slope of the absorption line, i.e., between the maxima of the derivative.) It is clear that the signal intensity will depend on the slope of the absorption line and the depth of the modulation. It should also be remembered that the shape of the signal will only be sinusoidal at small modulation amplitudes. With increase of the modulation amplitude the anharmonicity of the signal increases. We are only interested in the first harmonic of the signal (corresponding to the modulation frequency), which is the only harmonic passed by the receiver. It can be shown that the signal amplitude, i.e., the amplitude of this first harmonic, will be maximum when $U_m = 2\Delta f$, where $U_m$ is the modulation amplitude and $\Delta f$ is the half-width of the absorption line. In this case at any instant the signal amplitude is proportional to the corresponding ordinates of the absorption line, and the output signal describes the shape of the absorption line. (This last statement is not entirely accurate. During the passage through the maximum of the absorption curve the phase of the voltage changes, and an ordinary radio receiver is insensitive to changes in phase. Therefore the curve displayed on the oscilloscope screen represents the absolute value of the first derivative of the absorption line. A picture corresponding to the true character of the signal can be obtained only with the use of a phase-sensitive detector.)

It is not difficult to see that the high-frequency modulation of the signal generator, in which double modulation of the radiation source takes place, is completely analogous to the method of double modulation of the magnetic field used in observation of magnetic resonance spectra (Section 6, Figs. 16-18).

The advantages of the source-modulation method become especially apparent in the suppression of the false bursts. It is obvious that in this case, just as in the case of a spectrometer with a simple crystal-video detector, the pressure of the gas in the cell should be chosen so as to make the width of the absorption line smaller than the narrowest false burst. The high-frequency modulation in the region of these bursts is transformed to amplitude modulation in the same way as in the region of an absorption line. There are several possible cases.

1. If the depth of modulation exceeds the width of the burst, it will be fully observed on the oscilloscope screen together with the useful signal. In this case we have the same situation as in the case of a radio frequency spectrometer with a crystal-video detector. The false signal is decreased with the use of a filter that cuts off lower frequencies. This of course causes some distortion in the line shape, but it is smaller than in the previous case because now it is possible to decrease the amount of filtering on account of the improved sensitivity resulting from amplification at the modulation frequency.

2. If the modulation depth is smaller than the width of the burst but larger than the width of the actual line, a trace of the derivative of the burst envelope appears on the screen. The intensity of the parasitic signal in this case is drastically smaller so that the degree of filtering can be decreased still further, resulting in a decrease in distortion of the useful signal. When the modulation depth is decreased, the intensity of the parasitic signal decreases and the intensity of the useful signal remains constant. More precisely, the amplitude of the useful signal even increases (until the modulation amplitude becomes equal to the width of the line).

3. If the depth of modulation is smaller than the width of the absorption line, the useful signal is also observed in the form of the first derivative but the parasitic signal becomes very weak and also considerably broadened, which also enables an easier filtering of this signal. With this method a gain in sensitivity larger than a factor of 10 may be obtained relative

to the simple method of frequency sweeping. In particular, with small modulation and the use of a very narrow-band synchronous detector, a still larger gain in sensitivity may be obtained. The use of a synchronous detector becomes indispensable in the millimeter wavelength region since the absorption decreases with a decrease in power and the relative magnitude of reflection can increase at the same time.

The method of source modulation, together with phase-sensitive detection, is indispensable, indeed almost the only possible way to obtain high sensitivity in gaseous radio frequency spectrometers in the millimeter wavelength region. A block diagram of such a radio frequency spectrometer is shown in Fig. 63 (this radio frequency spectrometer is designed for the region of 2-3 mm; the method of frequency measurement is not shown). Here GHM is the high-frequency modulation generator

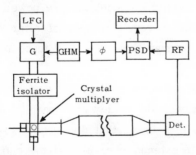

FIG. 63. Block diagram of a high-sensitivity spectrometer for the millimeter wavelength region

(4 kc/s) whose voltage is applied to the repeller of the klystron and is also used as the reference voltage in the phase-sensitive detector (PSD). The waveguide, which starts at the generator, terminates in the multiplier crystal which radiates the power into a smaller waveguide, which is smoothly coupled to the

waveguide of the absorption cell. From the detector crystal the signal is applied to a preamplifier (radio frequency amplifier) and after going through the synchronous detector is recorded by a chart recorder.

The second popular method for increasing the sensitivity of gaseous radio frequency spectrometers is the utilization of molecular modulation. The essence of this method is as follows. A special electrode is placed in the absorption cell and a radio frequency voltage which produces an alternating electric field within the gas volume is applied. Under the influence of the electric field, Stark splitting of the rotational levels takes place and the absorption line is either split or shifted. If at some instant the signal generator frequency coincides with the maximum of the absorption line, with application of the electric field, the absorption will decrease. In this way the alternating electric field amplitude-modulates the power incident on the detector, and as a result the signal after detection can be amplified by a narrow-band amplifier tuned to the modulation frequency.

The Stark absorption cell consists of a section of large-cross-section waveguide inside which, parallel to the broad face, is placed a strip of brass that divides the waveguide into two equal parts and extends over almost the entire length of the cell. The strip is fastened in place with polystyrene or Teflon supports. The introduction of the electrode into the cell is accompanied by well-known difficulties because it can strongly distort the high-frequency field and become a source of interference. A universal solution does not exist in this case, and a good Stark cell can be built only by long, hard experimental effort. The main requirements, however, reduce to clean mechanical processing of all parts, accurate location of the electrode in the center of the waveguide ($\pm 10 \ \mu$), an electrode mounting that is as rigid as possible in order to avoid vibrations, and tapering of the electrode at both ends (to decrease reflections).

The observed line shape depends strongly on the waveform of the modulation voltage. In the case of a square wave the obtained pattern is relatively simple and allows simultaneous study of the Stark splitting, meaning determination of the electric dipole moments of the molecules.

Let us assume that the voltage applied to the electrode is equal to zero during the first half of the cycle and to some positive value during the second half. If the period of the modulation voltage is smaller than the time required to sweep across the line (most often it is 1/5 of this time), during the zero-voltage half-cycle the absorption line itself will appear at the appropriate frequency, and during the second half cycle its Stark components will be observed. Therefore the observed pattern will be complicated because the line and its satellites will be observed simultaneously. The separation between the peaks depends, of course, on the amplitude of the modulation. The amplitude required for optimum observation depends on the line width (i.e., the pressure of the gas), the dipole moment of the molecule, and the type of Stark effect. For a linear Stark effect and molecules with an average dipole moment of the order of 1 debye and with a gas pressure of 10 n/m$^2$ ($\sim 10^{-1}$ mm Hg) a field strength of the order of 10 v/cm is sufficient. With sinusoidal modulation the pattern is usually much more complicated, but it can still be interpreted. Individual lines can be represented in terms of their first and second derivatives, accompanied by their satellites. Nevertheless, sinusoidal modulation is often used because it simplifies the design of the equipment.

The main advantage of Stark modulation, as far as sensitivity is concerned, is that it allows detection of absorption lines while preventing the detection of power changes due to reflections in the waveguide and, especially, in the absorption cell, even when the absorption line is broader than these parasitic signals. This is because the interferences are not modulated. Furthermore, it should be clear that this method, just as the preceding one, allows the use of amplification at high frequencies, thus cutting off the main portion of the noise spectrum.

Thus the method of Stark modulation allows one to work at higher gas densities, so that in turn the radiation density can be increased (in certain instances the power incident on the detector is on the order of several milliwatts), causing a corresponding improvement in the sensitivity of the radio frequency spectrometer. Moreover, there are a number of gas molecules with relatively broad lines even at very low pressures; these gases can therefore be investigated only by this method.

It would appear that the sensitivity of a spectrometer could be improved by increasing the modulation frequency and, consequently, the frequency at which the amplification takes place, since the noise is reduced with an increase in frequency. This, however, is not completely true. First, with an increase in frequency the capacitive reactance of the cell decreases, and consequently the required power from the modulation source increases excessively. Thus a compromise must be made which results in a decrease of the cell dimensions and a subsequent decrease in sensitivity. Second, experiments have shown that at very high modulation frequencies the lines are appreciably broadened.

In addition to the complexity of the observed spectrum, which often contains additional valuable information, the above spectrometers have a number of other faults. First, part of the sensitivity is lost because the lines are split into a large number of components and are thus less intense. Furthermore, the electrode plate located in the cell increases the mismatch between the cell and the source, and the electrode supports cause additional dielectric losses. All these factors also contribute to a decrease in sensitivity.

Stark modulation is preferred to source modulation because with Stark modulation the parasitic signals caused by reflections are not detected. Therefore the most widely used high-sensitivity gaseous radio frequency spectrometers are spectrometers with Stark modulation (with the exception of the millimeter wavelength region, in which difficulties are encountered in construction of the absorption cell). A simplified block diagram of a high-sensitivity Stark radio frequency spectrometer is shown in Fig. 64. A subject of particular concern is the stability of

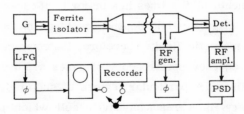

FIG. 64. High-sensitivity Stark spectrometer

the klystron. The klystron is shielded from temperature, vibrational, electric, magnetic and other perturbations. When a recorder is used for studying narrow lines, the frequency must be swept very slowly. Often a motor is used for this purpose, connected through a reducing gear to a mechanical tuning device. During search investigations, when an oscilloscope is used, the recorder is disconnected and, if previously a motor was used, an electric sweep is substituted in its place.

In addition to molecular modulation by the electric field, magnetic-field molecular modulation is possible for paramagnetic molecules (NO, $O_2$, $NO_2$, $ClO_2$ and others) and free radicals. In order to apply an alternating magnetic field, the absorption cell (a waveguide) is usually placed inside a solenoid. In order to eliminate eddy currents at high modulation frequencies a slot is made along the center of the broad face of the waveguide. For modulation frequencies below 1000 c/s this slot is usually omitted. In all other respects the performance of a spectrometer with Zeeman modulation is analogous to that of a Stark spectrometer.

To conclude this section, let us examine how the Zeeman effect is observed in rotational spectra of paramagnetic molecules. In gases one usually observes not direct transitions between Zeeman sublevels but splitting of the rotational levels. The components of the lines in a majority of cases do not have a sufficient magnetic field dependence to be modulated by the alternating magnetic field. Therefore, the common methods of the radio frequency spectroscopy of gases are usually used to detect the lines. A particular feature of such radio frequency spectrometers is a resonant-cavity cell which permits the application of a strong, homogeneous magnetic field to the gas. A high-$Q$ resonant cavity is used to increase the sensitivity, and the gas pressure is chosen to make the absorption line narrow in comparison with the resonance curve of the cavity. Further improvement in sensitivity is obtained by high-frequency modulation of the signal source.

## Chapter VIII

## Ammonia Beam Molecular Oscillator

### 28. THEORY OF THE MOLECULAR OSCILLATOR

A molecular oscillator using a beam of ammonia molecules was the first device utilizing the principle of microwave power generation by stimulated emission. The term "maser," which later became a universal name for numerous types of other devices working on the principle of induced radiation, including quantum paramagnetic amplifiers, was first used in conjunction with a molecular oscillator. In addition, an ammonia beam molecular oscillator basically differs from the masers described in Chapters V and VI. Paramagnetic masers are based on the phenomenon of electron paramagnetic resonance, while an ammonia beam molecular maser makes use of the transitions between the components of the inversion splitting of rotational levels. A necessary element of paramagnetic masers is the external magnetic field which establishes the sublevels, the transitions between which are used for induction of radiation. This induced emission results from the interaction of the magnetic moments of the atoms with the magnetic component of the amplified electromagnetic signal. The ammonia beam molecular maser, on the other hand, is based on the interaction of the electric dipole moment with the electric component of the electromagnetic field, and the induced radiation takes place as a result of transitions between the inversion sublevels, whose existence is independent of the presence of external magnetic or electric fields.

The above differences between paramagnetic masers and masers of the ammonia beam molecular oscillator type provide a reason for examining the operation of a molecular oscillator separately from paramagnetic masers.

The idea of molecular oscillation was first proposed in 1952 by Basov and Prokhorov and, independently, by Gordon, Zeiger and Townes. By 1954 the first ammonia beam molecular oscillators had been developed (USA and USSR). The 3—3 transition ($J = 3$, $K = 3$) with a frequency of 23,870 Mc/s was used in this maser. At present, molecular oscillators have been built in 10 countries.

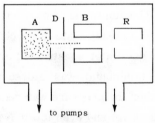

FIG. 65. Principal components of a
molecular oscillator

The construction of a molecular oscillator is shown schematically in Fig. 65. The oscillator consists of three main parts. First, it is necessary to have a source of the molecular beam A. The molecular beam should be such that the number of active molecules (i.e., the difference between the number of molecules in the upper and lower states) is positive; i.e., if possible, all molecules in the lower state should be removed from the beam. For this purpose the molecules are sorted with the help of the state-selector system B. Following this the molecular beam (with a positive number of active molecules) enters the resonant cavity R, which is tuned to the frequency of transitions used in the system. The feedback in a molecular oscillator is provided by the electromagnetic field of the cavity, which interacts with the dipole moments of the molecules that pass through it, inducing them to radiate. In order to obtain self-sustained oscillations, the resonant cavity should have a $Q$ that is sufficiently high to make the power radiated by the molecules greater than the power losses in the cavity and the oscillator load. The mechanism for starting the oscillations is as follows. Initially the molecules will radiate energy as a result of the effect of thermal radiation in the cavity, for example. The electromagnetic energy will be stored in the cavity and will again stimulate the active molecules entering the cavity to radiate. The amplitude of oscillations in the resonant cavity will grow until a steady state is reached.

In the molecular oscillator constructed by Basov and Prokhorov [60] the gas (ammonia) was stored in a high-pressure tank (20 atm), and then the pressure was reduced to 2 atm (with an oxygen reducing valve) and even lower (with a needle valve). After this the gas entered the vacuum system through a copper foil grill (holes $0.05 \times 0.05$ mm spaced 0.05 mm apart, diameter 5 mm). Such a grill allows one to obtain a beam of molecules with a limiting density of the order of $10^{18}$ molecules per second (corresponding to a beam pressure of the order of

1 mm Hg). A liquid-nitrogen-cooled collimator (aperture diameter 6 mm) was placed between the beam source and the system. This collimator decreased the number of molecules reflected back from various surfaces toward the beam source. Furthermore, the entire molecular oscillator was placed in a double-walled chamber, with the space between the walls filled with liquid nitrogen. This helped maintain a better vacuum because the molecules that hit the wall were trapped and their further movement within the vacuum system was thereby prevented.

The molecular beam obtained in this way contains molecules distributed among the energy levels, with a smaller number of molecules in high levels than in lower levels. Thus there is a problem of selecting the molecules; i.e., it is necessary to remove the molecules in the lower energy level of the transition of interest.

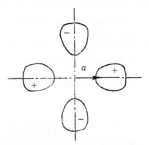

FIG. 66. Cross section of a
cylindrical quadrupole focuser

For a beam of ammonia molecules the most effective state-selecting method uses a cylindrical quadrupole focuser (Fig. 66). With the use of such a focuser a constant, strongly inhomogeneous electric field is obtained. The interaction of the ammonia molecules which travel along the axis of such a focuser and which are in different energy states (rotational as well as inversion) is described in terms of the quadratic Stark effect. The theory of this interaction is somewhat complex, and therefore we shall assume as a postulate the following conclusion of that theory: In an electric field the energy of the upper inversion level of the $NH_3$ molecule increases and the energy of the lower level decreases. Since every system tends to its minimum energy state, it follows from the preceding statement that the molecules in the upper energy level, when placed in an inhomogeneous electric field, will tend to move to the region of minimum $E$-field strength, where in this case the increase in energy due to the external electric field will tend to be minimum (Fig. 67).

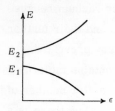

FIG. 67. Schematic representation of the Stark effect for the $NH_3$ molecule

On the other hand, since the energy of the molecules in the lower energy state decreases with the external electric field, these molecules will tend to move into the region of maximum electric field where the decrease in energy due to the external electric field will yield the minimum.

The deflecting force $F$ acting on the molecules in a quadrupole focuser is equal to

$$F = \pm \frac{p^2 K^2 M_J^2 V^2 r}{J^2 (J + 1)^2 h \nu a^4}$$

where $p$ is the dipole moment of the molecule; $K$ is the projection of the total angular momentum on the axis of the molecule; $M_J$ is the projection of the total angular momentum in the direction of the external field; $\nu$ is the frequency of the inversion transition; $V$ is the potential difference between the condenser plates; $a$ is the distance from the axis of the system to the surface of the electrodes; and $r$ is the distance of the molecule from the axis of the system.

Thus the molecules in the upper energy level which for some reason have deviated from the axis of the system will be returned to the axis by the force $F$. The amplitude of the oscillations about the axis will decay. Assuming a maximum possible deviation of a molecule from the axis at the end of the focuser, the required length of the focuser can be estimated.

In the molecular oscillator constructed by Basov and Prokhorov the quadrupole focuser had the following characteristics: length 10 cm, electrode gap 6 mm, operating voltage between electrodes $\sim$ 30 kv. A further increase in the number of active molecules due to an increase in the voltage was limited by electric breakdown of the condenser at 40 kv.

After passing through the state-selector system, the molecular beam with a positive number of active molecules enters the cavity which is tuned to the frequency of the spectral line. Here the molecules will be deexcited and will emit power. The emitted energy will be partially stored in the volume of the resonant cavity and partially lost in its walls. If the losses are smaller than the radiated power, the energy stored in the cavity will increase; i.e., the cavity will become excited and self-sustained oscillations will occur.

Neglecting the losses in the load, the condition for self-sustained oscillations of a molecular oscillator can be written as [69]

$$N_{act} h\nu > P$$

where $N_{act}$ is active number of molecules in the beam passing through the cavity per unit time and $P$ is the power lost in the walls of the cavity.

The power lost $P$ is related to the $Q$ of the resonant cavity. Expressing the loss $P$ in terms of $Q$ we obtain

$$N_{act} h\nu \;>\; 2\pi\nu\,\frac{W}{Q} \tag{107}$$

Assuming that the energy in the cavity is distributed uniformly over its volume we can write

$$W = \rho(\nu)V_0 \tag{108}$$

where $V_0$ is the cavity volume and $\rho(\nu)$ is the energy density such that

$$\rho(\nu) = \frac{3h^2(\Delta\nu)^2}{4\pi\,|d_{mn}|^2} \tag{109}$$

where $\Delta\nu$ is the half-width of the spectral line and $d_{mn}$ is the dipole moment matrix element of the molecule. Thus

$$N_{act} > \frac{2\pi V_0}{hQ}\;\frac{3h^2(\Delta\nu)^2}{4\pi\,|d_{mn}|^2} = \frac{3V_0 h(\Delta\nu)^2}{2Q\,|d_{mn}|^2} \tag{110}$$

From this relationship it is easy to find the required number of active molecules for any given $Q$ of the resonant cavity; or, vice versa, given the number of active molecules the required $Q$ of the cavity may be determined. The $Q$ is rarely smaller than $10^4$. For example, Basov and Prokhorov used a silver-plated cylindrical resonant cavity 20 mm long and 10 mm in diameter. The $Q$ of this cavity in the $E_{001}$ mode was $12 \times 10^3$, accounting for entrance and exit openings for the molecular beam (6 mm in diameter), waveguide coupling openings (3 mm in diameter) and the turning screw.

The frequency and amplitude of the oscillations set up in the resonant cavity are defined by the equation for the electric field intensity in a cavity*:

*Here and in the following we will denote by E the electric field strength.

$$\frac{d^2E}{dt^2} + \frac{\omega_0}{Q}\frac{dE}{dt} + \frac{\omega_0^2}{\epsilon}E = 0 \tag{111}$$

where $\omega_0$ is the resonant frequency of the cavity, $Q$ is its quality factor, and $\epsilon = \epsilon' + i\epsilon''$ is the complex dielectric constant of the substance filling the cavity.

Since the substance filling the cavity is the beam of ammonia molecules, which pass through the cavity in time $\tau$ and radiate at the frequency $\omega_l$, in general $\epsilon$ will be a function of the oscillation frequency in the cavity $\omega$, the electric field strength $E$, the spectral-line frequency $\omega_l$, the geometric dimensions of the cavity, and the transition probability. A detailed theoretical analysis leads to the following result:

$$\epsilon' = 1 - A\gamma \frac{(\omega_l - \omega)\tau}{(\omega_l - \omega)^2 + \frac{1}{\tau^2} + \gamma|E|^2}$$

$$\tag{112}$$

$$\epsilon'' = -A\gamma \frac{1}{(\omega_l - \omega)^2 + \frac{1}{\tau^2} + \gamma|E|^2}$$

where

$$A = \frac{2hN_{act}}{Sl}; \quad \gamma = \left|\frac{2\pi d_{mn}}{h}\right|^2 \tag{113}$$

($S$ is the cross-sectional area of a resonant cavity of length $l$, $|d_{mn}|^2$ is the square of the matrix element of the $z$ component of the dipole moment for the molecular transitions between states $m$ and $n$, $\tau = l/\overline{v}$ is the average transit time of the molecules through the field of the resonant cavity, and $\overline{v}$ is the average velocity of the molecules). The formulas for $\epsilon'$ and $\epsilon''$ were derived under the assumption that the field $E$ is uniform over the cross section of the resonant cavity. If we take into consideration the spatial dependence of the fields we do not obtain anything essentially new, and the problem can be reduced to the introduction of some effective field.

We seek the solution for $E$ in the form

$$E = E_0 e^{i\omega t}$$

Substituting this expression into Eq. (111) and equating the real and imaginary parts to zero we obtain

$$-\omega^2 + \frac{\epsilon' \omega_0^2}{\epsilon'^2 + \epsilon''^2} = 0$$

$$\frac{\omega}{Q} + \frac{\epsilon'' \omega_0}{\epsilon'^2 + \epsilon''^2} = 0$$

(114)

We now substitute into these equations the expressions for $\epsilon'$ and $\epsilon''$. After elimination of the amplitude of oscillations $E$ we obtain an equation for the steady-state frequency of oscillation:

$$\omega^3 + \omega \omega_0^2 \left( \frac{\omega_0 \tau}{Q} + \frac{1}{Q^2} - 1 \right) - \frac{\omega_0^3 \omega_l \tau}{Q} = 0$$

Assuming that $\delta = \dfrac{\omega_l - \omega_0}{\omega_l} \ll 1$ and solving the equation for $\omega$ (accurate to linear terms in $\delta$), we obtain the frequency of oscillation [61, 62]:

$$\omega = \omega_l \left( 1 + \frac{2Q}{\omega_0 \tau} \frac{\omega_0 - \omega_l}{\omega_l} - \frac{1}{Q \omega_l \tau} \right)$$

(115)

It is seen from this result that the frequency of oscillation is not equal to $\omega_l$, even when the resonant cavity is tuned exactly to the frequency of the spectral line ($\omega_l = \omega_0$). The difference between the frequencies $\omega$ and $\omega_l$ is very small. Basov and

Prokhorov give the following numerical values for a molecular oscillator operating on the $J = 3, K = 3$ line of $NH_3$ molecules:

$$Q \sim 1000, \quad \omega_0 \tau \sim 2 \times 10^7, \quad (\omega_0 - \omega_l)/\omega_l \sim 5 \times 10^{-6}$$
$$2Q/\omega_0 \tau \sim 10^{-4}, \quad 2Q\delta/\omega_0 \tau \sim 5 \times 10^{-10}$$
$$1/Q\omega_l\tau \sim 0.5 \times 10^{-10}$$

It is immediately evident from these figures that a molecular oscillator can be used as an absolute frequency standard with an accuracy of about $10^{-9}$.

The amplitude of the steady-state oscillations can be determined from Eqs. (114). For example, substituting into the second of these equations the values of $\epsilon'$ and $\epsilon''$ from (112) we obtain the value of $E_0^2$. Assuming that $\omega_0 \approx \omega_l \approx \omega$ and $Q^{-1} \ll 1$, we get

$$E_0^2 = \frac{A\gamma\tau^2 Q - 1}{\gamma\tau^2} \tag{116}$$

From this we obtain the second expression of the conditions for self-sustained oscillation:

$$E_0^2 > 0 \quad \text{or} \quad A\gamma\tau^2 Q > 1$$

i.e.,

$$\frac{2hN_{act}}{Sl} \left| \frac{2\pi d_{mn}}{h} \right|^2 Q\tau^2 > 1$$

or

$$\frac{8\pi^2 N_{act}}{V_0 h} |d_{mn}|^2 Q\tau^2 > 1 \tag{117}$$

If $A\gamma\tau^2 Q \gg 1$, then $E_0^2 \approx AQ$, i.e.,

$$E_0^2 = \frac{h^2 N_{act}}{\pi Sl} Q \tag{118}$$

The maximum possible power that can be obtained from a molecular oscillator is obviously equal to

$$P_{max} = \frac{1}{2} N_{act} h\nu$$

The molecular oscillator power obtained experimentally was of the order of $10^{-10}$ w (e.g., with a voltage of 32 kv on the focuser electrodes the power was $8.1 \times 10^{-10}$ w).

The difference between the frequency of oscillation and the frequency of the spectral line arises because the radiation of molecules along the cavity does not take place in a uniform manner and as a result, accompanying the primary mode of oscillation, with which the energy is propagated along the cavity, we also have a secondary mode of oscillation. This secondary mode of oscillation causes a Doppler shift in the frequency of the molecular transition. The effect can be compensated to a significant degree by letting the beam enter the cavity symmetrically on both sides.

Independent tuning of the resonant cavity to the frequency of the spectral line is a very complex problem because, first, the cavity is coupled to the waveguide system and, second, the substance filling the cavity in the form of the molecular beam also affects its resonance frequency. Therefore, there exists an entire "science" which deals with methods of tuning resonant cavities of molecular oscillators. The most convenient is the method which utilizes the frequency dependence of the oscillations on the intensity of the molecular beam. The limit to the accuracy of tuning of the resonance cavity is determined by the hyperfine structure and the nonuniformity of radiation of molecules along the length of the cavity.

A particular property of molecular oscillators is their extreme frequency stability because the frequency of oscillation is only weakly dependent on external factors and is mainly determined by the frequency of the utilized spectral line. Therefore a molecular oscillator can be used as a frequency standard, as well as a high-resolution radio frequency spectrometer. It is important to note that the bandwidth of a molecular oscillator is smaller than the bandwidth of the spectral line. Actually, initially, when the molecules radiate under the stimulation of noise, the radiation intensity is almost independent of frequency over the entire spectral line width; i.e., initially the radiation has a frequency spread equal to the width of the spectral line. Subsequently, however, a larger number of molecules will radiate at frequencies that are nearer the peak of the spectral line, where the density of radiation in the resonant cavity is higher, because the induced radiation can take place only at the frequency of the external source. Therefore the line will narrow down toward its peak. In the steady state, monochromatic oscillations will be established in the resonant cavity; their frequency will be close to the frequency of the peak of the spectral line.

If the conditions for self-sustained oscillations in a molecular oscillator are not satisfied, such a device can be used as a microwave power amplifier. A molecular amplifier is described by the same equation as a molecular oscillator, but a driving term, related to the power being amplified, must be added to the right-hand side of Eq. (111) [61]:

$$\frac{d^2E}{dt^2} + \frac{\omega_0}{Q}\frac{dE}{dt} + \frac{\omega_0^2}{\epsilon}E = \omega^2 B e^{i\omega t} \tag{119}$$

The amplitude of the external field $B$ is determined by the power $P_{in}$ supplied to the resonant cavity of the amplifier. Let us introduce the quantity $k = P_{out}/P_{in}$, where $P_{out}$ is the power extracted from the molecular amplifier. In the absence of the molecular beam the quantity $k$ is the power transfer efficiency through the resonant cavity. In order to find the gain, the field in the resonant cavity is computed for two cases: in the presence of the molecular beam (field $E_1$) and in its absence (field $E_2$). Then the power gain will be equal to the ratio of the square of the fields in the resonant cavity for these two cases, multiplied by the power transfer efficiency $k$ evaluated in the absence of the beam, i.e.,

$$k = k_0 \left(\frac{E_1}{E_2}\right)^2$$

The expression for $E_1$ can be obtained from the equation describing the operation of the molecular amplifier. The quantity $E_2$ can be easily found when $E_1$ is known. In fact, if $\epsilon' = 1$ and $\epsilon'' = 0$, then $E_1 \to E_2$.

At resonance, $\omega_0 = \omega_l = \omega$, the gain has its maximum value:

$$k_{max} = \frac{k_0}{A\gamma\tau^2 Q - 1} \tag{120}$$

The bandwidth of molecular amplifiers is relatively narrow and the tuning range is only several thousand cycles per second. Because of this the device has not found wide application as an amplifier.

## 29. PRACTICAL APPLICATIONS OF MOLECULAR OSCILLATORS

1. *Frequency comparison* [63, 64]. A molecular oscillator can serve as the basis for construction of frequency (and time) standards and for control of frequency of secondary standards.

Let us investigate how the frequency of a quartz oscillator can be determined with the help of a molecular oscillator. A block diagram of the main components of the system for frequency comparison of a quartz and a molecular oscillator is shown in Fig. 68. The frequency of the quartz oscillator QO is multiplied (first by a vacuum-tube mixer and then by a crystal diode) to the frequency of the molecular oscillator MO (for example, if the quartz oscillator is designed for 500 kc/s the total frequency multiplication will be of the order of 47,740). The harmonic obtained in this way should be sufficiently spectrally clean. For this purpose the tuning of the mixer circuits,

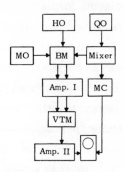

FIG. 68. System for comparison of frequencies of a quartz and a molecular oscillator [63, 64]

in order to prevent the appearance of phase modulation, must have an accuracy which is not achievable with ordinary tuning methods. Therefore the tuning must be done with the help of a spectrum analyzer, which analyzes the spectrum of the output signal. The output signal of the mixer (frequency $\nu_1$, power $10^{-8}$ w) is applied to one of the diodes of the balanced mixer BM. The balanced mixer also receives the signal from the molecular oscillator (frequency $\nu_2$) and the heterodyne signal (frequency $\nu_3$). A klystron serves as the heterodyne oscillator; its frequency is multiplied until it is equal to the molecular oscillator frequency, with the power of the heterodyne signal being of the order of $10^{-4}$ w. The balanced mixer forms intermediate frequencies ($\nu_2 - \nu_3$ and $\nu_1 - \nu_3$), which are amplified by an amplifier system and fed to a vacuum-tube mixer VTM, which in turn forms the frequency $\nu_2 - \nu_1$. (Instability of the heterodyne frequency is thereby eliminated.) From the tube mixer, after additional amplification, the signal is fed to the measuring circuit MC (to the horizontal deflection plates of an oscilloscope).

Through frequency division and addition, the measuring circuit forms the signal $\nu_2 - \nu_3$, which is then compared on the oscilloscope screen (applied to the vertical deflection plates on the oscilloscope) with the signal of the same frequency present on the horizontal plates. Thus the oscilloscope screen will show an ellipse. From the rate of rotation of this ellipse the frequency stability can be estimated. Revolution of the ellipse 1-2 times per second corresponds to a relative stability of the quartz oscillator (heterodyned up in frequency) of the order of $10^{-9}$-$10^{-10}$ in a time interval of several seconds.

The voltage output of a signal generator can be used in place of the measuring circuit. This will give rise to Lissajous figures on the oscilloscope screen, and the rate of rotation of these figures will correspond to the frequency stability. If the quality factor of the quartz is low ($Q < 10^5$-$10^6$), a sharp Lissajous figure may not appear on the screen. The relative instability of the multiplied frequency will be of the order of $10^{-8}$.

2. *Frequency stabilization* [65]. The power output of a molecular oscillator (MO) is very small ($\sim 10^{-10}$ w). The lack of efficient amplifiers in the MO region ($\lambda \sim 1.25$ cm) prevents the design of an oscillator (by conventional techniques) with substantially larger power output and frequency stability comparable to that of a molecular oscillator. It is possible, however, to use a method of automatically phase locking the frequency (APL) of a stabilized oscillator, in which an MO provides the reference signal (Fig. 69). The stabilized oscillator can be a

K-12 klystron ($P \sim 60$ mw), whose eighth harmonic has a frequency corresponding to the MO frequency. The eighth-harmonic oscillations of the stabilized K-12 klystron and the MO signal are fed to two superheterodyne amplifiers BM-I and BM-II, which are balanced mixers with a common heterodyne frequency (e.g., a K-18 klystron whose frequency is multiplied by three). The intermediate frequency of both amplifiers is the same (65 Mc/s).

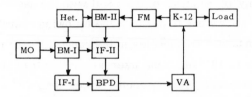

FIG. 69. Frequency stabilization system using a molecular oscillator [65]

The output signals from both IF amplifiers are fed to a balanced phase detector (BPD) whose output signal depends on the difference in phase of the two input voltages. This signal is amplified by a video amplifier and applied to the reflector of the stabilized K-12 klystron.

Since the amplifier IF-II is a part of the servo loop, its bandwidth should be as large as possible, to enable the APL system to operate in a broad frequency band. The amplifier IF-I is not a part of the control loop and its bandwidth should be as small as possible in order to decrease the noise level in the amplification circuit of the MO signal (the limit here is imposed by the frequency stability of the heterodyne signal).

3. *Molecular clock* [66]. It was indicated above that a molecular oscillator can be used as an absolute frequency (time) standard, with an accuracy of the order of $10^{-9}$. The relative

frequency stability of two molecular oscillators is even higher and is of the order of $10^{-11}$ (in a time interval of about 20 minutes). This refers to molecular oscillators operating on the $J = 3$, $K = 3$ line of a beam of $NH_3$ molecules and a frequency of about 23,870 Mc/s. The frequency of molecular oscillators can be tuned to the frequency of the spectral line with an accuracy of 10-20 c/s.

The molecular clock constructed by the Physics Institute of the Academy of Sciences of the USSR with the use of molecular oscillators allows one to perform the following operations:

1) To measure frequencies in the range $10^4$-$10^7$ c/s with an accuracy up to $10^{-9}$ in a time interval smaller than 100 seconds. For measurement of frequencies lower than $10^4$ c/s with an accuracy of $10^{-9}$, the measurement time must be increased.

2) To measure time intervals with an accuracy of $2 \times 10^{-7}$ seconds.

3) To provide frequencies of $5 \times 10^5$ c/s and its multiples with a relative stability of $10^{-9}$.

The circuit uses a highly stable quartz oscillator whose frequency is controlled by the frequency of a molecular oscillator (this circuit is used because of the difficulties encountered in the amplification of power levels of $10^{-9}$-$10^{-10}$ w and the direct division of the frequency of 24,000 Mc/s, which are obtained from the molecular oscillator).

Thus with the use of an APL system the frequency stability of the quartz oscillator is assured to be not worse than $10^{-7}$. This means that the period of oscillation $T$ is known with the same accuracy ($\sim 10^{-7}$). The stable frequency signal is fed to a very complicated system which forms pulses, whose repetition rate is controlled by a special timing system. The separation (in time) between the pulses will be $T$ (or some multiple of $T$). These pulses can be fed to a graphic recorder. Thus time markers, spaced by an interval $t$, can be produced on the recorder chart (obviously the recorder chart speed must be stable to an accuracy not worse than $10^{-7}$). If some other process (e.g., the passage through a magnetic resonance line) is simultaneously recorded on the recorder chart, the time of this process can be determined with an accuracy not less than $10^{-7}$.

# References

1. V. K. Arkad'yev (W. Arkadieff), Zh. Russ. Fiz.-Khim. Obshchestvo, Ser. Fiz., 1931, 45, 312.
2. A. Einstein, Z. Phys., 1922, 11, 31.
3. J. Dorfmann, ibid., 1923, 17, 98.
4. C. E. Cleeton and N. H. Williams, Phys. Rev., 1934, 45, 234.
5. L. D. Landau and E. M. Lifshitz, Phys. Zeits. Sowjunion, 1935, 8, 153.
6. C. J. Gorter, Paramagnetic Relaxation (translation from English), IL, Moscow, 1949 (original: Elsevier, 1947).
7. J. M. B. Kellog and S. Millman, Usp. Fiz. Nauk, 1948, 34, 72.
8. Shifts in Levels of Atomic Electrons (collection of articles translated from English), IL, Moscow, 1950.
9. E. R. Andrew, Nuclear Magnetic Resonance (translation from English), IL, Moscow, 1957 (original: Cambridge University Press, 1955).
10. J. D. Roberts, Nuclear Magnetic Resonance; Applications to Organic Chemistry (translation from English), IL, Moscow, 1961 (original: McGraw-Hill, 1959).
11. N. N. Neprimerov, Izvestiya AN SSSR, Ser. Fiz., 1954, 18, 360.
12. W. Gordy, W. V. Smith and R. Trambarulo, Microwave Spectroscopy (translation from English), GITTL, Moscow, 1956 (original: John Wiley, 1953).
13. D. J. E. Ingram, Spectroscopy at Radio and Microwave Frequencies (translation from English), IL, Moscow, 1959 (original: Butterworths, 1955).
14. S. A. Al'tshuler and B. M. Kozyrev, Electron Paramagnetic Resonance, Fizmatgiz, Moscow, 1961; English translation

(by Scripta Technica, Inc.), Academic Press, New York, 1964.

15. A. A. Manenkov and A. M. Prokhorov, Radiotekhnika i Electronika, 1956, 1, 469.

16. N. Bloembergen, E. M. Purcell and R. V. Pound, Phys. Rev., 1948, 73, 679.

17. A. V. Kubarev and Yu. A. Mezenev, Pribory i Tekh. Eksper., 1960, No. 2, 86.

18. D. J. E. Ingram, Free Radicals as Studied by Electron Spin Resonance (translation from English), IL, Moscow, 1961 (original: Butterworths, 1958).

19. A. G. Semenov and N. N. Bubnov, Pribory i Tekh. Eksper., 1959, No. 1, 92.

20. Ye. K. Zavoyskiy, Zh. Eksp. Teor. Fiz., 1945, 15, 344.

21. S. G. Salikhov, ibid., 1947, 17, 1070.

22. B. M. Kozyrev, Izvestiya AN SSSR, Ser. Fiz., 1962, 16, 533.

23. Ye. K. Zavoyskiy, Zh. Eksp. Teor. Fiz., 1947, 17, 155.

24. N. Bloembergen, E. M. Purcell and R. V. Pound, Nature, 1947, 160, 475.

25. B. R. Rollin and J. Hatton, Phys. Rev., 1948, 74, 346.

26. V. V. Lemanov, Pribory i Tekh. Eksper., 1961, No. 1, 126.

27. V. F. Bystrov, L. L. Dekabrun, Yu. N. Kil'yanov, A. U. Stepanyants and E. Z. Utyanskaya, ibid., 1961, No. 1, 122.

28. A. G. Semenov, ibid., 1962, No. 5, 5.

29. Yu. N. Denisov, ibid., 1960, No. 1, 82.

30. A. V. Kubarev, ibid., 1957, No. 3, 57.

31. Ya. I. Korol'kov and N. A. Bugrov, ibid., 1960, No. 2, 99.

32. N. M. Pomerantsev and V. I. Klivlidze, ibid., 1957, No. 2, 56.

33. N. M. Pomerantsev, Usp. Fiz. Nauk, 1958, 65, 87.

34. N. J. Hopkins, Rev. Sci. Instr., 1949, 20, 401.

35. R. V. Pound, Progr. Nucl. Phys., 1952, 2, 21.

36. L. Malling, Electronics, 1954, 27, 134.

37. E. I. Fedin, Vestnik AN SSSR, 1958, No. 7, 79.

38. L. A. Blyumenfel'd and V. V. Voyevodskiy, Usp. Fiz. Nauk, 1959, 68, 31; Vestnik AN SSSR, 1959, No. 12, 16.

39. H. Y. Carr and E. M. Purcell, Phys. Rev., 1954, 94, 630.

40. J. R. Singer, Masers (translation from English), IL, Moscow, 1961 (original: John Wiley, 1959).

41. G. Troup, Masers; Microwave Amplification and Oscillation by Stimulated Emission (translation from English), IL, Moscow, 1961 (original: Methuen, 1959).

42. J. P. Witke, Collected Volume: Millimeter and Submillimeter Waves (translation from English), IL, Moscow, 1959, pp. 252-296 (original: Proc. I.R.E., 1957, 45, 291).
43. F. E. Goodwin, Proc. I.R.E., 1960, 48, 113.
44. G. Feher, J. P. Gordon, E. Buehler, E. A. Gere and C. D. Thurmond, Phys. Rev., 1958, 109, 221.
45. P. F. Chester, P. E. Wagner and J. G. Castle, Jr., Phys. Rev., 1958, 110, 281.
46. J. S. Gooden, Radio Engineering and Electronic Physics Abroad (translation), 1959, Nos. 2, 3.
47. H. M. Goldenberg, D. Kleppner and N. F. Ramsey, Phys. Rev. Letters, 1960, 5, 361.
48. J. W. Meyer, Electronics, 1960, 33, 58.
49. Quantum Paramagnetic Amplifiers (collection of articles translated from English), IL, Moscow, 1961.
50. A. L. McWhorter and J. W. Meyer, Phys. Rev., 1958, 109, 312.
51. D. L. Carter, J. Appl. Phys., 1961, 32, 2541.
52. R. W. DeGrasse, D. G. Hogg, E. A. Ohm and H. E. D. Scovil, J. Appl. Phys., 1959, 30, 2013.
53. A. Blandin, Onde electr., 1961, 41, 931.
54. W. Gordy, Usp. Fiz. Nauk, 1949, 39, 201.
55. C. H. Townes and A. L. Schawlow, Microwave Spectroscopy (translation from English), IL, Moscow, 1959 (original: McGraw-Hill, 1955).
56. I. I. Gol'dman and V. D. Krivchenkov, Collection of Problems in Quantum Mechanics, GITTL, Moscow, 1957 (translated into English as Problems of Quantum Mechanics, Addison Wesley, 1961).
57. V. L. Ginzburg, Usp. Fiz. Nauk, 1947, 31, 320.
58. M. W. P. Strandberg, Microwave Spectroscopy (translation from English), IL, Moscow, 1959 (original: Methuen, 1954).
59. N. G. Basov, Radiotekhnika i Electronika, 1956, 1, 752.
60. N. G. Basov, Pribory i Tekh. Eksper., 1957, No. 1, 71.
61. N. G. Basov and A. M. Prokhorov, Usp. Fiz. Nauk, 1955, 57, 485.
62. N. G. Basov and A. M. Prokhorov, Zh. Eksp. Teor. Fiz., 1956, 30, 560.
63. G. A. Vasneva, V. V. Grigor'yants, M. Ye. Zhabotinskiy, D. N. Klyshko, Yu. N. Sverdlov and Ye. I. Sverchkov, Izvestiya Vysshikh Uchebnykh Zavedenii, Fizika, 1958, 1, 185.
64. V. V. Nikitin, ibid., 1958, 1, 1960.

65. I. L. Bershteyn, Yu. A. Dryagin and V. L. Sibiryakov, ibid., 1959, 2, 130.
66. N. G. Basov, I. D. Murin, A. P. Petrov, A. M. Prokhorov and I. V. Shtranikh, ibid., 1958, 1, 50.
67. N. G. Basov and A. M. Prokhorov, Zh. Eksp. Teor. Fiz., 1955, 28, 249.
68. F. Bloch, Usp. Fiz. Nauk, 1955, 56, 429.
69. N. G. Basov and A. M. Prokhorov, Zh. Eksp. Teor. Fiz., 1954, 27, 431.

# Index

226